Livestock Production in New Zealand

Livestock Production in New Zealand

EDITED BY KEVIN STAFFORD

Contents

Introduction

Kevin Stafford

Introduction

Kevin Stafford
Institute of Veterinary, Animal and Biomedical Sciences
Massey University, Palmerston North

Food is essential for life, and food production is a necessary process. Humans are omnivorous; the majority of people eat a range of plant-based and animal-based foodstuffs. While nations try to maintain food security, either by producing food locally or building trade relations with countries that produce excess food, few are self-sufficient in food. Indeed, New Zealand imports a lot of food both for humans and for animals. Nevertheless, in New Zealand livestock farming is a major industry that produces food for both national consumption and export.

One image of New Zealand is that of a pastoral scene, with green pastures grazed by sheep and cattle. This is an attractive image, but there are underlying tensions between livestock farmers and their critics. The latter worry about the impact of livestock on the environment, and may be concerned about animal welfare and the safety of food produced on New Zealand farms.

The growth of dairy farming, particularly in the South Island, has led to concerns about the long-term impact of irrigation on water resources and the negative effects of cow urine and faeces, plus fertiliser, on the quality of the surface water in streams and rivers and on drinking water. A growing concern for the welfare of farm animals has led to regular exposés of animal abuse, especially in the pig and poultry industries, but more recently also in the dairy industry. To this is added a growing lack of understanding of livestock production among the general population, as is to be expected with reduced rural populations and a concomitant growth in urbanisation. This lack of understanding may lead to poorer communication between urban and rural communities. A recent illustration of this occurred when some commentators on calf rearing did not appear to understand that to get milk from cows, their calves had to be weaned soon after

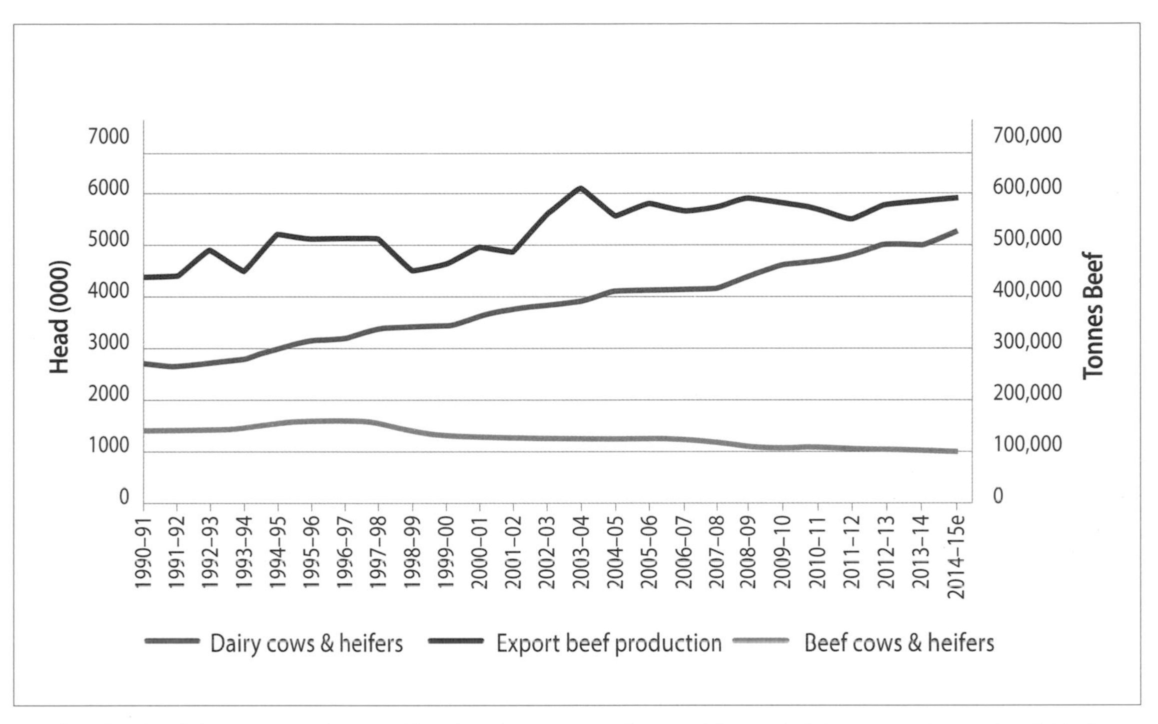

Trends in beef and dairy cow numbers, and beef production, over the past 25 years ('e' denotes estimates for 2014–15). Source: Beef & Lamb New Zealand Economic Service.

birth and be fed artificial replacement milk so that the cows' milk could be taken for human consumption. This is an old practice, obvious to farming folk but not to some of those unfamiliar with dairy farming.

The world has a large and increasing appetite for milk and meat. Global meat production and consumption over the past 50 years has trebled, to 312 million tonnes in 2014. The cost of food to the consumer is an important political issue; in most developed countries, people spend less than 13 per cent of their disposable income on food. The pressure to keep food prices low results in competition at the retail end of the market, and the prices paid to farmers for their produce are often only just above — or even below — the cost of production. This is clearly illustrated by the price that dairy farmers are currently being paid for milk (although farmer debt is a factor in this particular dilemma). The economic pressures on farmers caused by low prices for their produce are countered in the United States and Europe by subsidies that keep farming viable. In New Zealand, farming is not subsidised and many livestock farmers survive economically by keeping large herds or flocks and using low-input systems to maintain profitability. Today there is little scope for increased livestock numbers in New Zealand, given that land development has largely been completed.

New Zealand is known internationally both as a tourist destination and as a source of food. The country's economy depends on these two industries, of which food production is the more important. A significant proportion of export growth in the past decade has been from agriculture. The foods produced for export include milk products, meat (beef, lamb and venison), fruit, wine, honey and some vegetables. Poultry meat, eggs and pig meat are produced for national consumption. Minor livestock products include velvet, goat and sheep milk, and goat meat; horse meat is also exported. A multitude of other livestock products are exported, including wool, hides, tallow and meat meal. However, although New Zealand exports a wealth of livestock products, it also imports a lot of food — about 50 per cent of the food eaten here is imported, including lamb, beef and milk products and more than half of the pig meat consumed by New Zealanders.

Some important factors underpin New Zealand's position in world meat and milk trade. One is the disease-free status of its livestock. Foot and mouth disease (FMD) and bovine spongiform encephalopathy (BSE, commonly known as mad cow disease) are not present in the country, allowing it to export fresh chilled and frozen beef and lamb to a great number of markets. Nearly all of New Zealand's cattle and sheep are grass-fed, which has a particular advantage in some markets in the United States and Asia. Meanwhile, there are threats to New Zealand milk and meat exports from other countries. Some South American countries would be able to compete for our lamb and beef markets if they improved their animal health status, and the United States and the EU countries compete with us on the international milk market. Moreover, there is the potential for self-inflicted damage to our export markets. Modern communication can quickly disseminate news internationally and, as a 1080 incident in 2015 showed, it is important to be prepared to counter bad publicity immediately (Brackenridge, 2016). Being 'clean, green and kind' is an important marketing strategy for New Zealand food exporters — but it has to be real, as evidence to the contrary will be publicised by special-interest groups and perhaps even our market competitors. Attaining the goal of significantly increasing primary sector exports to NZ$64 billion by 2025 not only requires that care be taken to maintain our reputation (Peterson, 2016), but must also have public support.

This book has been written to inform people about how animals, including horses and dogs, are managed on farms in New Zealand. It has not been written to defend livestock production, nor to condemn it. It is simply an introduction to how animals are farmed commercially in New Zealand. Debate about how we manage land, water and animals in this country is ongoing, and this is a healthy aspect of our democracy. Informed debate is valuable, while uninformed debate is pointless, and probably damaging. This book was written to inform this debate.

References

Brackenridge, J. (2016). Why are we wasting a good crisis? The value shift our primary sector needs. In C. Massey (Ed.), *The New Zealand Land and Food Annual*, vol. 1 (pp. 27–36). Auckland: Massey University Press.

Petersen, M. (2016). Never say die: The way forward for the New Zealand agrifood sector. In C. Massey (Ed.), *The New Zealand Land and Food Annual*, vol. 1 (pp. 37–43). Auckland: Massey University Press.

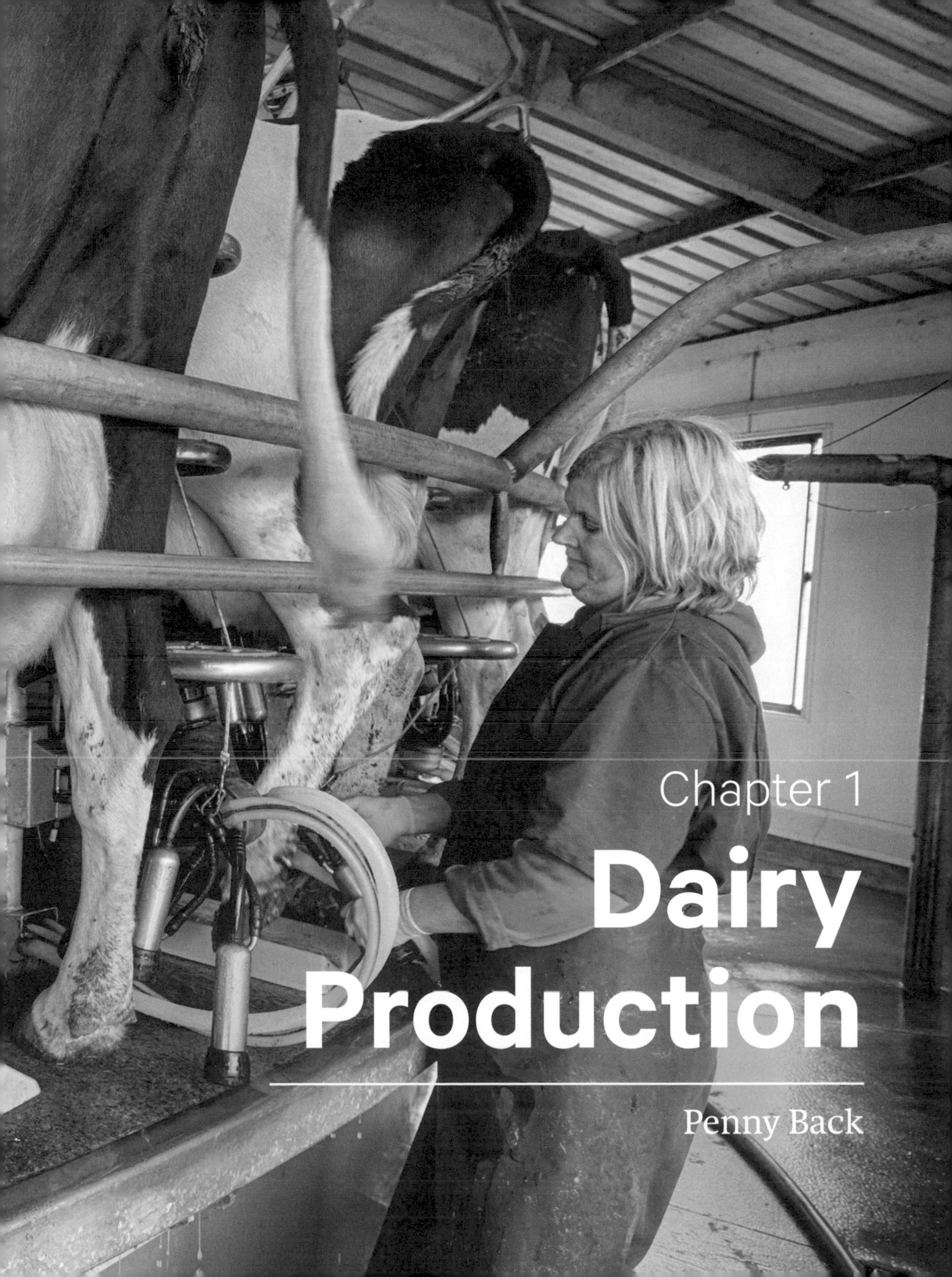

Chapter 1

Dairy Production

Penny Back

Chapter 1

Dairy Production

Penny Back

Institute of Veterinary, Animal and Biomedical Sciences
Massey University, Palmerston North

Introduction

In New Zealand, dairy production is based on the efficient conversion of grazed pasture into milk. Grazing pasture as a main source of feed is one of the main differences between New Zealand and the majority of the dairy systems in North America and Europe. The northern hemisphere system is more intensive — cows are typically housed and fed grains, concentrates, food industry by-products and conserved forage (silage), with little or no pasture. While these systems can produce greater milk yields per cow, they are more labour-intensive and have far greater running costs. Conversely, one of the New Zealand industry's strengths has traditionally been converting low-cost feed (pasture) into product, with a focus on milk production per hectare (Holmes et al., 2007). More recently, however, the focus has been on increasing milk production per cow through increased supplementary feeding of purchased feeds (such as those used in the northern hemisphere systems). This change has been driven by high milk prices, which make it financially feasible to use more expensive feeds. Increased milk production per cow is also desirable from a sustainability perspective: it is envisaged that in the future more milk will be produced from fewer cows, reducing the environmental footprint of dairy farming. However, importing more feed has increased the costs of production (cost per kg of milksolids produced), which on many farms cannot be supported with the current lower milk prices, and farmers are now focusing again on grazing management.

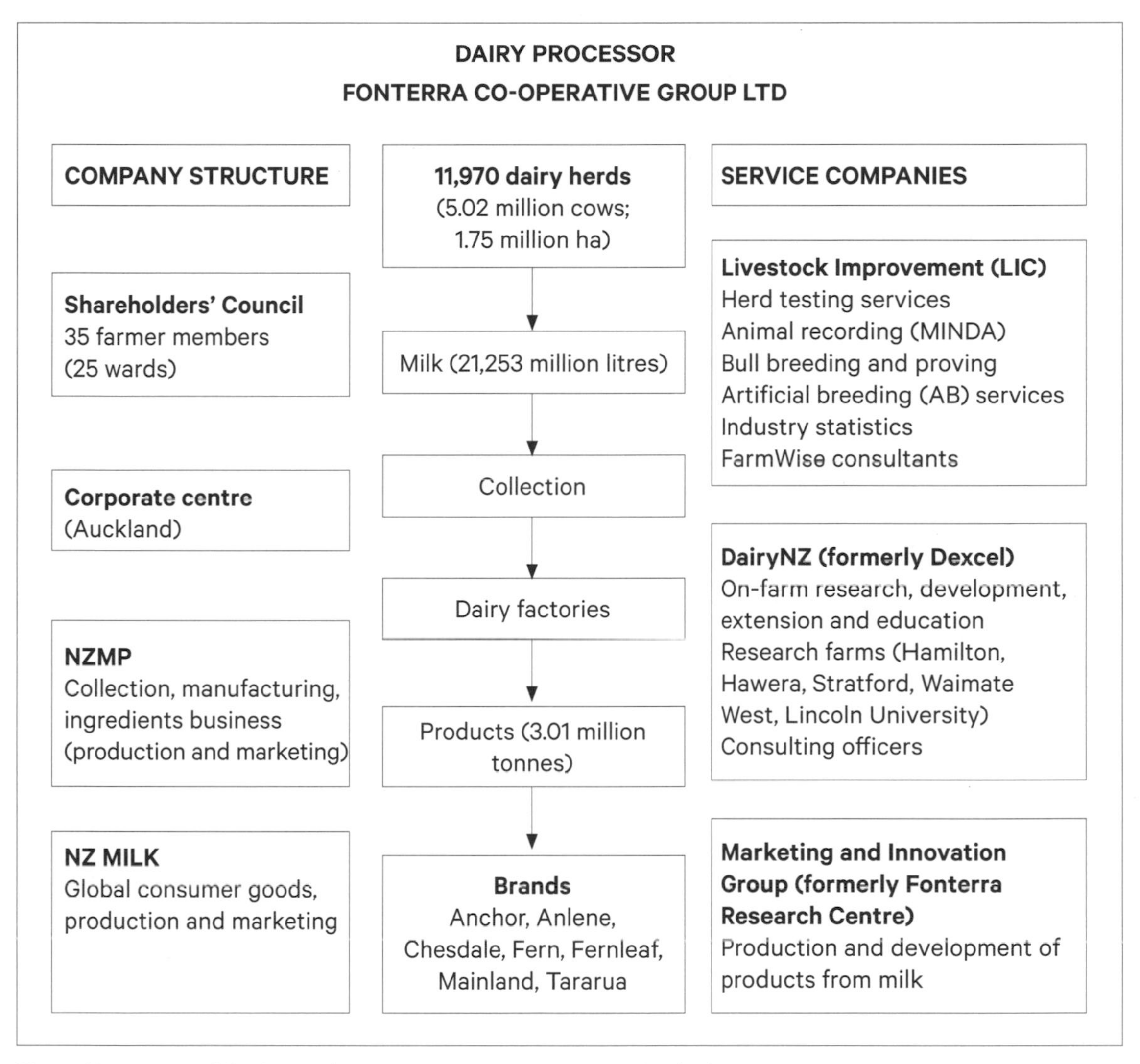

Figure 1.1 Structure of the dairy industry, using Fonterra as an example of a dairy processor.

The New Zealand dairy industry is an integrated system, encompassing the production, manufacturing and marketing of milk products. It is supported by companies with a focus on improving herd inputs. Genetics and herd recording are carried out by the LIC (Livestock Improvement Corporation) along with research, extension and advice from DairyNZ (Figure 1.1). Historically, the dairy industry was based on a co-operative, farmer-owned structure whereby milk processors focused on maximising financial returns to shareholder farmers; companies such as Fonterra, Westland and Tatua are co-operatives. Recently, some new corporate companies such as Open Country Dairy and Synlait have been purchasing milk on a contract basis. This has reduced the requirement for farmers to own shares and is starting to change the ownership structure of the industry.

The dairy season in New Zealand is seen as running from June to May each year, with payments to farmers based on kilograms of milksolids (kg MS); milksolids are the combined yield of fat and protein in the milk. The payment formula is based on fat yield plus protein yield minus milk volume. There are price penalties for failing to comply with specific quality specifications, for example for high somatic cell count (SCC) or the presence of antibiotics in the milk.

Dairy farmers are not paid for all of the milk they produce as they supply it. Prior to the start of each season (late May), a forecasted payout (price per kg MS) is announced; farmers are paid about 65–70 per cent of this forecasted payout during the milking season, until about March. The forecast is revised several times over the season if market signals require it. The final payout is decided the following October; meanwhile, the payments to farmers are finalised over the four months following the end of June. As the forecasted payout is revised over the season, budgets need to be routinely updated.

A new facet of the dairy industry in New Zealand is the recent growth in the goat and sheep dairy sectors. The Dairy Goat Co-op (www.dgc.co.nz), based in Hamilton, produces nutritional powder products (e.g. infant formulas and milk powder) and long-life milk from goats. Sheep dairying primarily involves cheese production, e.g. Whitestone cheese in Oamaru (www.whitestonecheese.co.nz). There are about 45 boutique cheesemakers in New Zealand, making a range of cow, buffalo, goat and sheep cheeses (New Zealand Specialist Cheesemakers Association, 2016).

New Zealand produces about 3 per cent of total world milk production, approximately 21 billion litres of milk per annum. It is a small producer compared with the USA, which produces 12 per cent of total world production. However, in contrast to most other countries, over 95 per cent of the milk produced in New Zealand is manufactured and exported, meaning that New Zealand accounts for approximately 30 per cent of all dairy products traded internationally. A wide variety of products is manufactured, including butter, cheese, milk powders, infant formulas, casein and protein products (Table 1.1).

Table 1.1 **Volumes (in '000 tonnes) of milk products produced**

	2001–02	2009–10	2012–13	2013–14	2014–15
WMP	nr	1568	1768	1694	2025
SMP	nr	472	585	614	534
Milk powders (total: WMP + SMP)	816	2040	2353	2308	2559
Cheese	342	373	459	452	490
Cream products	408	nr	nr	nr	nr
Butter	nr	217	242	245	248
Casein, protein products and albumins	157	nr	nr	nr	nr
Other dairy products	24	35	224	229	208
Total dairy products	1747	2665	3277	3234	3506

Source: Fonterra.co.nz.

WMP = whole milk powder; SMP = skimmed milk powder; nr = not individually reported for that season.

Dairy cattle on winter feed crop in Southland. **Previous:** Milking cows on a rotary milking platform.

In contrast to dairy industries in other countries, the New Zealand industry produces and exports without the support of government subsidies or export incentives. Fonterra sells milk products by auction, this having been introduced to bring transparency to international milk prices and the volumes of product sold. Fonterra was the innovator of this concept, with other international processors now starting to follow their lead.

The development of functional foods, such as low-fat, high-calcium and high-protein milk, has been a growth sector within New Zealand's dairy industry. Other areas that have been developed include biomedical and health products, such as colostrum-based supplements and hyper-immune milk products. Products for inclusion in sports drinks have been developed to aid the post-event recovery of athletes.

Currently, Fonterra controls around 85 per cent of the raw cows' milk produced in New Zealand and is the major dairy processor. The co-operative exports milk products to 151 countries, with increasing amounts going to 'developing' countries. Fonterra is the world's largest exporter of dairy products. Of the milk produced, 97 per cent is processed into product, with the remainder supplying the local liquid milk market and producers of other fresh dairy products for local consumption. Dairy exports were worth around NZ$14 billion in 2014–15 (around 8 per cent of GDP) to the New Zealand economy (Ministry for Primary Industries, 2016).

Glossary

DairyNZ: the industry organisation that represents New Zealand dairy farmers. It is funded by farmers through a levy payment, currently 3.5 cents for every kilogram of milksolids supplied to a milk processor. The purpose of DairyNZ is to 'secure and enhance the profitability, sustainability and competitiveness of New Zealand dairy farming'. Its activities include research, extension through consulting officers and the development of resources for farmers. For more information see: www.dairynz.co.nz.

Dry cow: a non-lactating cow.

Fonterra: the largest dairy processor in New Zealand. It is a co-operative that is owned by around 10,500 dairy farmers, and a global business. Fonterra processes 22 billion litres of milk a year worldwide; 18 billion litres are produced in New Zealand, with the remainder produced by local suppliers in Australia, Chile, Brazil, Sri Lanka and North America. Fonterra also owns two farms in China, with three more under development, and aims to have each producing 150 million litres a year.

Heifer: this term is used by farmers in a variety of ways. It may describe a young female cow that has not produced a calf or has had one calf, whereas heifer calves are calves aged from 0 to 9 months, and are then commonly referred to as R1s, meaning rising (nearly) one-year-old. They are referred to as R2 heifers after about 18 months of age. R2s that have calved are commonly called heifers for their first lactation.

LIC (Livestock Improvement Corporation): this organisation became a user-owned co-operative in 2002. Its activities include the provision of genetics, information and advice for dairy herd improvement. Services provided to farmers include herd testing, herd recording and artificial breeding, DNA analysis, research to improve farm profitability, farm automation, farm advisory services and statistical information relating to the New Zealand dairy industry. For further information see: www.lic.co.nz.

Milk solids: milk is 87 per cent water; the solids are what is left after the water is removed, i.e. fat, protein, lactose (sugar), vitamins and minerals. The milksolids (MS) measurement that is used to determine the payout to farmers is different, comprising just fat plus protein.

Oestrus: the sexualy receptive period of the sexual cycle.

Payout: the financial payment received by dairy farmers for milk supplied, based on kilograms of milksolids (kg MS).

Ruminant stomach: cows' stomachs are composed of four compartments — the reticulum, rumen, omasum and abomasum.

SCC (somatic cell count): this measures the number of somatic cells that are present in milk. The SCC increases rapidly when cows have an infection, and is thus used to measure milk quality.

Season: a dairy farming season or year runs from 1 June to 31 May, and events occur regularly at the same time every year.

A brief history of the dairy industry

The first cows were imported to New Zealand in 1815 by the Reverend Samuel Marsden, when Shorthorn cattle (a dual-purpose breed, used to produce both milk and meat) were shipped from Sydney to Russell, in the Bay of Islands. During the next 30 years increasing numbers were imported, so that by the 1840s small amounts of butter and cheese, mainly from Shorthorn cows, were being exported to Australia. In 1848, Ayrshire cattle were imported by Scottish settlers in Otago; and in 1862, one Jersey bull and two Jersey cows were imported to Whanganui. The first Friesian cattle were imported from the Netherlands in 1884 (Holmes et al., 2007). Thus, the parent breeds of the modern national herd were established early in the industry's development.

Dairy companies in New Zealand have traditionally always had a co-operative structure, where the farmers who supply the milk are also the company's shareholders and owners. The Otago Peninsula Co-operative Cheese Factory Co. Ltd was the first commercial dairy company, established in 1871. Exporting was a challenge initially, as products were carried by sea, but by 1882 refrigeration had been developed. By 1895, 218 dairy factories were in operation, and refrigeration had allowed new products such as condensed milk, casein and lactose to be added to the main products of butter and cheese.

The quality of the milk and dairy products became the focus. Following the development of a test for measuring the concentration of milk fat, a payment system based on the quantity of milk fat supplied was devised in the 1920s, when the number of dairy companies and factories reached its maximum with about 540 companies operating (Holmes et al., 2007).

Mechanical machine milking started in 1918; this enabled farmers to milk more cows, so herd sizes started to increase. The number of herds increased until 1920 and then decreased again, although the total number of cows and the number of cows per herd have both continued to increase since then. Milk yield per cow has also continued to increase with improved genetics, pastures and feeding. Seasonal fluctuations are evident due to the impact of factors that cannot be controlled, such as bad weather. These include droughts in summer or cold, wet springs, which can limit pasture supply and hence the amount of milksolids produced.

The breed composition of the national herd has changed over time. In the 1890s the majority of cows were Shorthorns; however, due to a focus on producing milk fat, the number of Shorthorns then decreased while Jersey numbers increased rapidly, so that by the 1960s Jerseys were the dominant breed. After 1965 the number of Friesian cows increased rapidly, dominating breed proportions due to calves being in demand for bull beef production.

Until the 1990s the production of milk fat had been the focus of measurement, payment and genetic selection; then routine measurement of the protein concentration in milk resulted in a change in the payment system. The concentrations of both fat and protein are now measured, and yields calculated and combined. The combined yield is expressed as 'milksolids'. Another major change resulted from the development and introduction of artificial breeding

(AB) in the 1950s. This allowed the genes from high-producing animals to be spread faster and wider in the national herd. More recently there has been a focus on crossbreeding, which has led to Holstein Friesian x Jersey crossbred cows (known as KiwiCross cows) becoming the dominant breed (LIC & DairyNZ, 2015).

The Dairy Board, which was responsible for marketing and selling all milk products, was established in 1923. It existed in several forms until 2001, when the two largest dairy companies (New Zealand Dairy Group and Kiwi Co-operative Dairies) merged with the New Zealand Dairy Board to form the Fonterra Co-operative Group Ltd.

A Friesian dairy cow.

Production systems in the dairy industry

Most cows do not lactate (produce milk) all year; the average lactation length, or days in milk, for cows in New Zealand is around 270 days. Lactation length is influenced by management and cow factors that include the age of the cow, its health status, the level of feeding (quality and quantity), milking frequency and calving date. The main factors affecting productivity on a dairy farm can be seen in Figure 1.2.

The challenge for farmers is to match feed supply with feed demand. Feed supply can be classified as the amount of pasture and other feed sources available to feed animals on farm. Feed demand is the animals' own requirements for feed. Demand is the sum of the energy required for maintenance of body function and for growth, pregnancy and lactation; the amount of energy for these processes differs depending

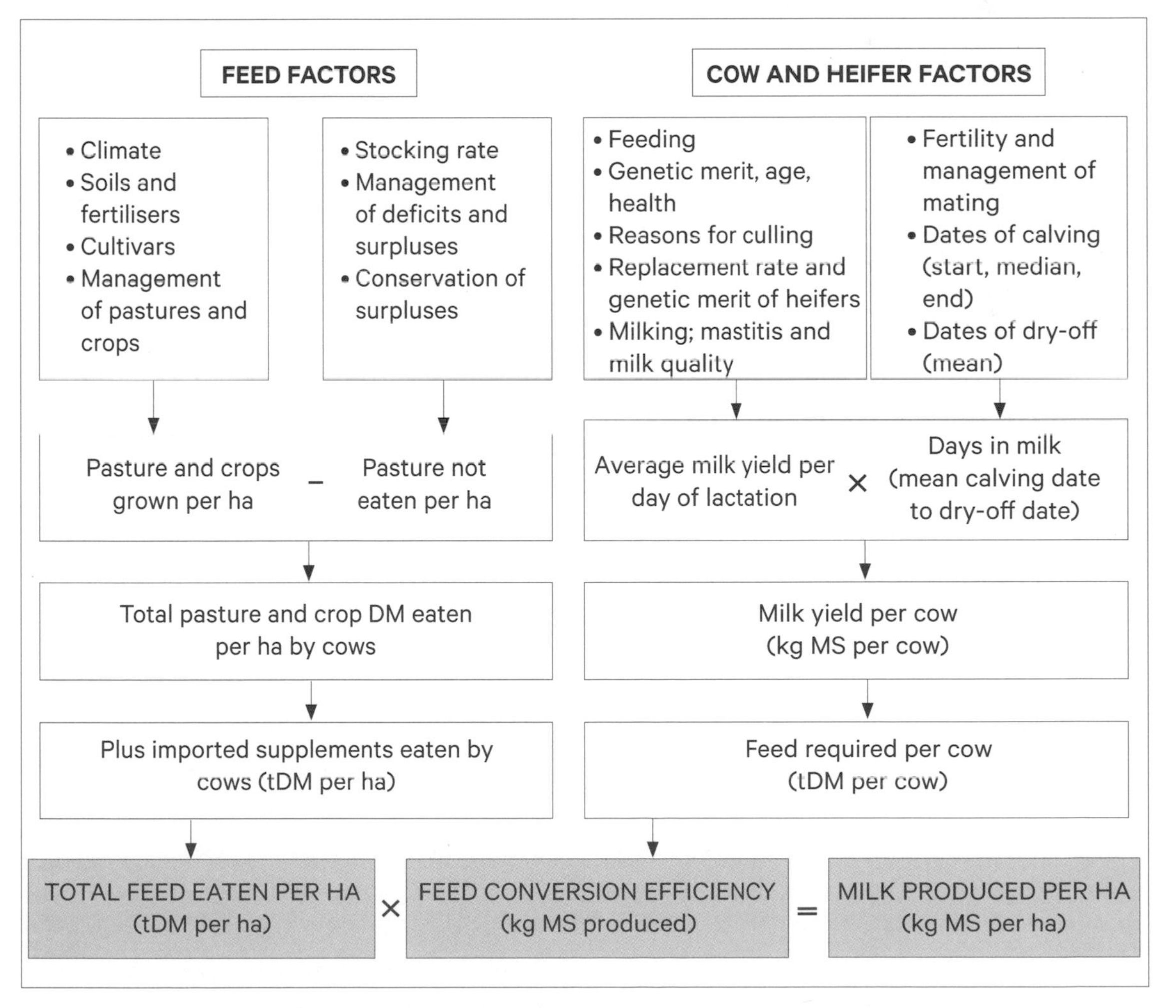

Figure 1.2 The main factors affecting productivity on a dairy farm, and how they combine. DM = dry matter; ha = hectare; MS = milksolids (fat + protein); tDM = tonnes of dry matter.

Table 1.2 **Feed intake, milk volume and milksolids production at different stages of lactation**

Stage of lactation	Feed intake (kg DM/ day)	Daily milk volume (litres)	Milksolids (kg MS/ day)
Early	15–18	20–25	1.6–2.0
Mid	14–16	15–20	1.2–1.6
Late	13–15	10–15	0.8–1.2
Dry	8–12	–	–

DM = dry matter; MS = milksolids (fat + protein).
Source: adapted from DairyNZ (2010).

on age, liveweight and stage of the season, i.e. whether a cow is lactating or not, and what stage of lactation she is at (Table 1.2). Detailed information on how this can be calculated can be found in Rattray et al. (2007).

In most areas of New Zealand, the basic diet for dairy animals is pasture that usually comprises ryegrass and white clover species. Some areas, such as Northland, use subtropical kikuyu pasture, which better tolerates the climate in that area. In periods when pasture growth exceeds animal demand, pasture is conserved as hay and silage and fed in times of feed deficit (summer, winter).

Different forages and crops can also be grown to provide extra feed for times when pasture does not grow well, such as dry summer months and colder winter months. These crops include pastoral herbs such as chicory and plantain, legumes such as red clover and lucerne, brassica crops such as turnips and kale, fodder beet, and cereal crops such as winter oats. Maize is grown either to make maize silage or to be fed as greenfeed maize, which can be grazed.

These crops not only provide more feed (quantity), but also provide feed of greater quality. Extra feed can also be purchased and bought on farm, such as cereal grains and blends. Palm kernel expeller (PKE), a by-product of palm oil production, is imported from Indonesia and Malaysia specifically as cow feed.

The challenge of matching feed supply with feed demand means that lactation lengths are shorter in pasture-based systems than the housed systems of the northern hemisphere, unless large amounts of supplementary feed are incorporated into the system.

The recent high milk prices resulted in many farmers building concrete feed pads and winter housing barns to supply cows with extra feed during lactation and to extend lactation into winter. While this increased feeding increases milk production, it also increases costs. With lower milk prices, less supplemental feed is supplied in this way and stocking rates are reduced on many farms.

Following the recent change in feeding systems with increasing amounts of supplements being imported on to farms, DairyNZ has described five production systems so that farmers can compare their farm's performance with others using a similar system. These definitions are based primarily on when imported feed is fed to dry stock or lactating cows during the season. It also includes off-farm dry cow grazing, but does not include grazing or feed for young stock.

- **System 1:** all grass, self-contained, all stock on farm (traditional system). No feed is imported on to the farm, and no supplement is fed to the herd except for that harvested on farm (grass or maize silage). Dry cows are not grazed off.
- **System 2:** approximately 4–14 per cent of total feed is imported, either as supplement or in the form of dry cow grazing off farm.
- **System 3:** approximately 10–20 per cent of

Three Friesian calves and a beef cross calf (white head).

total feed is imported to extend lactation in autumn and for dry cows. Farms feeding 1–2 kg of meal or grain per cow per day for most of the season best fit into this system. The highest proportion of farms in New Zealand operate under this system.

- **System 4:** approximately 20–30 per cent of total feed is imported and used at both ends of lactation and for dry cows.
- **System 5:** approximately 25–40 per cent of total feed is imported and used throughout lactation and for dry cows. This system also includes farms where more than 40 per cent of cow diet is sourced from off farm.

Seasonal milk production — spring calving

Pasture growth is much greater in spring than in winter, and the majority of herds in New Zealand calve in late winter/early spring (July to September) to produce milk during spring, summer and autumn. This enables herd feed requirements to be matched with pasture growth rates. It is important for seasonal milk production to maintain a 365-day calving interval, so cows must become pregnant again between October and December to ensure that they will calve again in the following July to September. The gestation period of cows is about nine months.

Following spring calving, milk production declines slowly over summer and autumn, and cows are dried off once milk yield drops to around 5–6 litres per day or the next calving date is approaching. Drying off means that farmers stop milking the cows and over the next few days they stop lactating. This happens in late summer/

Housed dairy cattle eating total mixed rations — a method of feeding which combines forages, grains, proteins and minerals into a single feed mix.

autumn (March to May), and depends on pasture growth, milk yield and other factors (body condition, days to next calving). All spring-calving cows are dried off before winter so that their dry period (when they don't produce milk and their feed requirements are lowest) coincides with the period when pasture growth is lowest.

In New Zealand, unlike Europe or North America, the majority of cows remain outside throughout the year. They graze and spread dung and urine. These two features of farming make for a cheaper farming system, as feeding indoors and managing cow excrement and urine add large costs in terms of housing, machinery, labour and feed costs in more intensive systems. However, the situation is starting to change; dairying has moved into areas such as Southland that have heavy soils with a high water-holding capacity and a more difficult climate in the winter for farming cows. Housing is being introduced there, particularly during the winter months, as having the animals off wet, muddy paddocks is good not only for pasture protection and animal welfare but also for managing effluent and run-off into surface water. Housing must be well constructed, with sufficient room for the number of animals housed, stalls (cubicles) of the correct depth to allow cows to lie down comfortably and suitable access to food and water.

Winter milk systems

A portion of the national herd calves in autumn so as to be milked during winter to supply fresh milk and associated products to the domestic market. These farms have a 'winter milk contract' with their dairy company, which specifies that a certain volume of milk must be supplied daily during a specific period (e.g. May to July). Dairy companies use winter milk contracts to make better use of their plant and to ensure continual supply of product. Farmers with a winter milk contract receive a premium price for milk supplied during the contract period.

To supply milk during winter, cows calve in March and April (autumn) and are mated in June to July (winter). They are dried off in December or January (summer). On most farms with a winter contract there will be two herds, an autumn-calving herd and a spring-calving herd. The spring herd calves 'as normal' in spring, and these farms therefore supply milk for 365 days a year. In some regions, for example parts of Northland, that have warm winters and dry summers, some herds calve entirely in autumn and produce winter milk only, to avoid trying to milk through the very dry summers experienced in those regions.

Once-a-day milking

For cows to keep producing milk, the milk has to be removed either by a calf or through milking. Over the past century, cows have been selected to perform well when being milked twice daily. Once-a-day (OAD) milking was used as a short-term strategy to overcome feed shortages such as those experienced during a dry summer, or in adverse weather conditions such as floods. More recently, however, some farmers have started milking OAD for the whole season. This system can be of benefit in many situations, such as long, thin farms or those with steep hills where the cows are walking significant distances. On some farms it is possible to milk an increased number of cows by having two herds, one that is milked in the morning and the other in the afternoon, enabling an increase in herd size without having to invest money in building a new milking shed.

By reducing milking to OAD, a lower milk yield is expected. While this indeed happens to some cows, others are not greatly affected. Selection for cows that tolerate the reduced milking frequency has led to farmers whose herds have been milked under this system for an extended period of time reporting little effect on overall milk production (Mikkelsen, 2012). However, achieving this result takes three to four seasons and farmers must be in a financial position that enables them to cope with reduced milk income during that time. Management of mastitis (infection of the udder) also needs to be a top priority with cows that are only being monitored OAD at milking. However, established OAD herds usually have lower SCCs and no increase in clinical mastitis cases. Cows with high SCCs or that have had recurring cases of mastitis should be culled before starting OAD milking.

There are also beneficial effects in terms of reproduction and the incidence of lameness. With a reduced milking frequency, cows are better able to maintain their body condition; this results in being able to identify cows that are cycling more easily, and in better conception rates and six-week in-calf rates. Improved reproduction has resulted in farmers having greater numbers of heifer calves to choose replacements from (Mikkelson, 2012). The surplus heifers are often reared and then sold, which creates a new income stream; hence, farm profitability may not be significantly reduced. Other advantages can be a requirement for fewer staff or lower staff turnover, and a better lifestyle. Less time spent milking cows means that farmers can be more involved in family, community and industry events.

Pasture and herd management

The seasonal, pasture-grazed system is possible

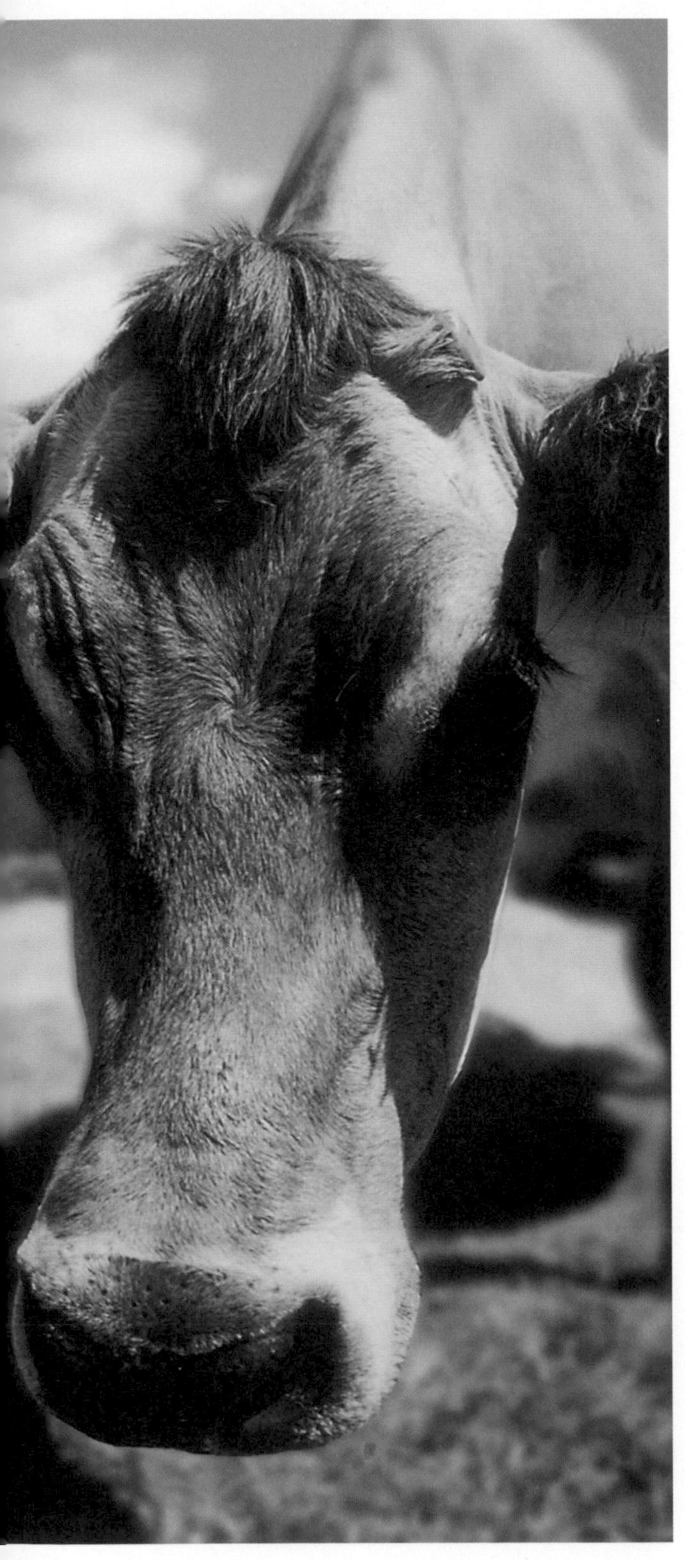
A Jersey cow.

in New Zealand due to the physical and temperate climatic conditions that allow pasture production to continue all year round in many regions. The main dairy areas in the North Island are capable of growing at least 10 tonnes of dry matter per hectare (tDM/ha) per year — this requires at least 1000 millimetres of rain per annum and an altitude of up to 400 metres above sea level. In drier areas, grass production can reach 20–22 tDM/ha if irrigation is used. Pasture production is affected most strongly by climate and soil fertility, and ranges between 8 and 16 tDM/ha annually in the main dairy areas (see Table 1.3).

Pasture quality can be maintained by having pasture and clover in a leafy state, which supplies around 10–12 megajoules of metabolisable energy per kilogram of dry matter (MJ ME/kg DM) and 15–30 per cent crude protein. Older pasture with a high proportion of old leaves and stems results in a lower feeding value, around 6.5 MJ ME/kg DM. The differences in pasture production between areas is one of the main causes of differences

Table 1.3 **Examples of pasture production across New Zealand (north to south)**

Location	Annual pasture production (tDM/ha/year)
North Island	
Dargaville	17.0
Otorohanga	16.0
Stratford	12.3
Massey (No. 4 Farm; wet clay)	11.3
South Island	
Nelson	13.4
Lincoln (light sandy and heavy clay)	17.6 and 14.2
Edendale	12.8

Source: adapted from DairyNZ (2010).

Table 1.4 **Average sizes and production of cows in different regions of New Zealand**

District	Average herd size	Cows grazed per ha	MS produced per cow per year (kg)	MS produced per ha (kg)
Northland	311	2.28	317	722
Waikato	335	2.97	370	1099
Taranaki	291	2.85	395	1124
Wairarapa	366	2.77	360	997
Westland	414	2.20	346	760
North Canterbury	808	3.50	416	1457
Southland	590	2.77	381	1055

Source: adapted from LIC & DairyNZ (2015). ha = hectare; MS = milksolids (fat + protein).

in milksolids production (Table 1.4); however, the use of irrigation and of greater amounts of bought-in feed now play a large part in milksolids production. The amount of pasture grown, and how effectively it is harvested and converted into milk by cows (utilisation), is influenced by many management factors, such as stocking rate, calving and dry-off dates, fertiliser use, intensity of grazing patterns, drainage and fencing (Holmes et al., 2007).

The stocking rate is expressed as the number of adult cows being grazed per hectare. Stocking rates are important as they determine the amount of pasture that is eaten by cows. High stocking rates utilise virtually all of the spring pasture growth but cause a deficit in available pasture in summer. This then requires the feeding of supplements and/or earlier drying off to prevent cows losing body condition and becoming thin. In contrast, low stocking rates can result in pasture being wasted in spring if it is not eaten, and in loss of pasture quality. Farmers must determine the optimum stocking rate for their farm by balancing all of these factors.

In spring-calving systems, the choice of calving date, the date the herd starts calving, depends on pasture growth rates in late winter/early spring, the availability of other feeds (e.g. silage), the use of nitrogen (N) fertiliser and the stocking rate. Calving dates vary between regions (see Table 1.5). Dry-off date is chosen so as to reduce herd feed requirements and to balance cow needs with low winter pasture growth rates. Dry cows do not need to eat as much feed as milking cows.

The dry-off date is also chosen to allow cows enough time to gain weight and regain body condition before calving again. Cows are generally dried off when milk production is approximately 5 litres or less. They require a dry period of at least 60 days to allow the mammary gland to shrink and repair before the next lactation. If cows are milked for too long, they can become thin and may then still be thin when they calve the following spring; this will decrease milk yield and reproductive performance in the next season (i.e. they will produce less milk and take longer to get in calf).

The 'average' New Zealand dairy farm

Table 1.5 gives data for an 'average' dairy farm in New Zealand.

For most dairy farms, the milking platform (area of the farm used for the milk herd) is supplemented by extra land for grazing off heifers and dry cows over winter and by purchased-in feeds. Yields of milk per cow are relatively low compared with overseas, due to shorter lactation lengths, lower feed intakes, smaller cow sizes and the high stocking rates required to harvest pasture effectively. There are differences between the North and the South islands in terms of herd and farm size. Average herd sizes are increasing in both islands, but South Island herd sizes and farm sizes are increasing faster than in the North Island. In 2014–15, there were 2 million cows in the South Island compared with 3 million in the North Island. The largest average herd size in the South Island is 808 cows (North Canterbury) compared with 652 cows in the North Island (Hawke's Bay). The smallest average herd sizes are found in the North Island: Auckland (272 cows), Taranaki (291 cows) and Northland (311 cows) (LIC & DairyNZ, 2015).

South Island farms have higher herd production (in terms of total MS); for example, in 2014–15 North Canterbury was highest with 355,847 kg MS as a result of larger herd sizes, high stocking rates and high MS produced per cow. In the North Island, Hawke's Bay was highest with 233,909 kg MS due to large herd sizes. Average production per hectare and per cow were higher in the South Island (1182 kg and 390 kg, respectively) than in the North Island (1022 kg and 368 kg) (LIC & DairyNZ, 2015).

The high cost of dairying land plus higher financial returns compared with sheep and beef in recent years has seen the expansion of dairying into non-traditional areas such as Canterbury and Southland. The availability of irrigation in Canterbury allows increased pasture growth rates and therefore greater production.

Dairy calves (Friesian and KiwiCross).

Table 1.5 **Data for the 'average' seasonal-supply dairy farm and 'average' cow for 2014–15**

Cows	419 milking cows (includes about 92 heifers which are first-calving two-year-olds)
Area	146 ha (ignoring races, plantations, etc.)
Stocking rate	2.87 cows/ha
Young stock	100 heifer calves 93 one-year-old heifers (yearlings)
Feed	Growing 10–16 t pasture DM/ha Consuming 10–13 t pasture DM/ha Main fertiliser inputs: 25–50 kg P/ha; 20–100 kg N/ha plus some feed 'imported' and some dry stock (heifers and cows) grazed away from the home farm
Labour	Operated by about two people with help from casual labour at busy times such as calving
Ownership	67% are owner-operators 32% are sharemilkers
Production	213 kg milk fat/cow (4.78% fat) 164 kg protein/cow (3.84% protein) 377 kg milksolids/cow
Breeds	34.7% Holstein Friesian (HF) 10.4% Jersey (J) 45.6% HF x J crossbreds 0.7% Ayrshire 8.6% other (e.g. Brown Swiss, Guernsey)
Breeding	Breeding worth (BW) = $106 73% of herds artificially inseminated
Calving date	Early July for upper North Island Mid to late July for lower North Island Early August for South Island
Herd testing	73% of cows are herd tested (3.65 million cows)

Source: adapted from LIC & DairyNZ (2015). ha = hectare; kg N/ha = kilograms of nitrogen per hectare; kg P/ha = kilograms of phosphorus per hectare; t pasture DM/ha = tonnes of pasture dry matter per hectare.

However, this has come at an increased cost of capital investment, meaning an increased cost of production; these farms must therefore be high-producing and efficiently run to dilute this cost. In Southland, the challenge of heavy soils and more severe winters has resulted in increased use of housing. It remains to be seen whether the increases previously described continue with the forecasted lower payouts over the coming seasons.

Calendar of dairy farming operations

LATE WINTER TO EARLY SPRING
(JULY TO OCTOBER)

Calving

Most herds calve in spring, and calves are removed from their mothers within 24 hours of birth. On most dairy farms, newly calved cows and their calves are brought in from the paddock and separated once a day. Calves should be tagged and navel-sprayed (to prevent infection) in the paddock. The identity of the mother of the calf is recorded, as this is needed to establish each calf's breeding worth (BW) and breed make-up. Each calf is placed in the calf shed with other calves of the same age. The cow joins the colostrum cow mob to start milking.

Calf management

On most dairy farms, a proportion of the female calves (heifers) are kept to be reared as replacements for cows that are sold (culled) from the herd. Unwanted heifer or bull (male) calves are sold, either to the bobby calf trade (for slaughter as veal), or at calf sales where they are purchased by commercial calf-rearers. A small proportion (usually Jersey bull calves) may be reared as breeding bulls. These bulls are leased to other farmers to get dairy cows in calf and then are either retained or sent to slaughter. A proportion of calves are born very small or weak and unhealthy, and may be killed on farm. These are unwanted as it is unlikely that they will grow into worthwhile cattle. Calves should be killed using a captive bolt gun followed by pithing (destruction of brain tissue) and perhaps bleeding out. Blunt force is not recommended for killing calves.

When a cow calves, the milk she initially produces is called colostrum. It is different from normal milk and is not supplied directly for human consumption. Farmers cannot legally supply milk from the first eight milkings after calving, unless contracted to do otherwise, so the colostrum and early milk are normally stored and fed to calves. It is also used in the manufacture of human health products for improved immune function, healing injuries, improving muscle mass and aiding sports recovery. It is not used for the usual milk products (cheese, butter, milk powder). The composition of colostrum changes quickly to that of milk over a few days.

Colostrum is very important for the newborn calf. The calf has a naïve immune system, and the mother's colostrum supplies the antibodies needed to protect the calf from disease for the first few months of its life. It is therefore important that calves receive colostrum to prevent illness. Calves are managed so that they receive first-milking colostrum, which should have high levels of antibodies, during their first feeds in the sheds. The calves have to be taught to drink from feeders with rubber teats. If they don't drink, they can be stomach-tubed to ensure that they receive colostrum. Calves should receive at least 2 litres of first-milking colostrum, preferably 10 per cent of their body weight (up to 4 litres) soon after they are moved into the calf shed.

Bobby calves can be sent for slaughter at four days of age. They must be a minimum of 20 kg liveweight. They must have been fed before transport, be able to support their own weight, have a dry navel and be healthy. They should have received the same level of care as the calves that are being held on the farm or being sold at calf sales.

When a calf is born, it is essentially monogastric — milk is digested and absorbed through the abomasum (the fourth stomach compartment) alone. The rumen (the major stomach compartment) starts to develop when the calf begins to nibble grass (if it is in a paddock) or hay/concentrates. There are many different calf-rearing systems; they vary in the amount of milk that is fed, the types of feeds offered, how many times a day the calf is fed, the number of calves in a pen, and how quickly the calves are put outside. In general, on most farms, when calves are small they are fed twice a day with a minimum of 4 litres per day of milk. A hard feed, such as concentrates or calf meal/pellets, is also offered in the calf pens so that the calves will start to nibble on it and will recognise it in the paddock. Calves may also be offered some hay. The purpose of offering these different types of feed is that this gives another way of getting energy into young animals so that the amount of milk being fed can be reduced, and it also helps with rumen development.

Calves spend their first few weeks in the calf shed and are then put outside to graze, the timing depending on the size of the shed, the number of calves it can house and the region. The number of calves kept in each pen is important — if there is overcrowding, this increases the risk of disease. Bedding (e.g. sawdust, wood shavings) must be kept clean and dry, and draughts should be prevented. Calves need to be warm and dry; they should use the food they consume for growth, not for warmth or fighting infections. Calves can go outside from a young age, but must be in sheltered paddocks to reduce the risk of diseases caused by exposure, like pneumonia. If calves are wet and cold they will not grow, and in a worst-case scenario, they will die. A sheltered paddock with good-quality pasture is much better for them than being kept inside in overcrowded, wet, dirty pens.

Weaning off milk and concentrate feeds will generally occur between eight and 12 weeks of age, and calves should be approximately 20 per cent of mature body weight at this time. There are established growth rates and target liveweights for replacement heifer calves to meet (Table 1.6). It is important that calves are big enough to have reached puberty by 12 months of age, so that they can have several oestrus cycles before being mated at 15 months of age. Successful breeding at 15 months is required for the heifer to calve at 24 months of age, and to

Table 1.6 **Liveweight targets (kg) with daily intake requirements (kg DM) given in brackets**

	Age				Average mature liveweight (kg)
	8–12 weeks	6 months	15 months (mating)	22–24 months (calving)	
% mature weight	20%	30%	60%	90%	–
Jersey	80 (2.7)	120 (3.5)	240 (5.7)	360 (8.4)	400
HF x J	90 (3.2)	135 (4.1)	270 (6.6)	405 (9.9)	450
Holstein Friesian	100 (3.6)	150 (4.6)	300 (7.6)	450 (11.2)	500

Source: adapted from DairyNZ (2010). HF x J = Holstein Friesian x Jersey cross (KiwiCross).

Cows being milked on a rotary platform.

then have time to recover and get in calf again while lactating to match the seasonal calving pattern of the herd.

Target growth rates from weaning to calving are 0.4–0.8 kg/day (an average 0.66 kg/day). It is recommended that growth be 0.7–0.8 kg/day before puberty, and 0.5 kg/day for 90 days before calving to ensure a body condition score between 5 and 5.5 by the time of calving (see page 42 for more on body condition scores). These figures are estimates for animals grazing pasture of 11 MJ ME/kg DM.

Getting replacement heifers to reach target liveweights during summer and winter has been identified by the dairy industry as an area that needs attention. Data analysed by the LIC (McNaughton & Lopdell, 2012) shows that weight gains during early rearing are good, but calves start to fall behind target liveweights during their first summer and most never recover, so that 75 per cent are under target liveweights by the time they calve at two years of age.

Milking

The mammary gland of a cow (the udder) has four individual (separate) quarters, each with a teat. Milk must be removed through the teats regularly for lactation to continue. Cows are usually milked twice a day (TAD) between 4 a.m. and 9 a.m. and again between 2 p.m. and 6 p.m. This gives intervals between the start of milkings of approximately 10 hours during the day and 14 hours during the night. During this time cows graze and ruminate during resting. Cows need to spend about 40 per cent of their time resting.

Cows in New Zealand are milked predominantly through two types of dairy shed: a herringbone or a rotary platform. A herringbone has two lines of cows either side of a sunken pit down the middle. There may be up to 100 cows on each side, with a set of cups for each cow. Cows enter one side and are milked while the other side empties and refills. The cows come in at one end and exit from the other. On a rotary platform, cows move onto a moving circular platform one at a time, and exit when the platform has moved through 300 degrees or so. The number of cup clusters available to milk the cows depends on the size of the rotary platform — there can be up to 120 bales (framework for securing the cows), although more commonly there are about 50–60.

A cluster of four cups is placed on each cow; one cup per teat. The cup has a rubber liner and its pulsation mimics the suckling of a calf. Milk is released (let down) from the udder through the teat. It is transported by a vacuum in the milk line to a storage vat, where it is cooled to 4°C and maintained at that temperature. A tanker from the milk processor usually picks up the milk from the vat once a day, during either the day or the night between milkings.

Frequent removal of milk is needed for continued milk production, but care must be taken not to over-milk or under-milk — this can lead to udder health problems (mastitis, cracked teats or sore teats). Once the cow is milked and the cups removed, the teats are sprayed with a teat spray to protect the udder from infection. This spray contains disinfectant to kill bacteria and an emollient to promote teat skin health. Milking machines are cleaned after every milking and must be well maintained to minimise mastitis and harvest milk efficiently. This ensures the safe collection of high-quality milk, prevents damage to the teats and udder and prevents the transmission of bacteria between cows. The transmission of bacteria from dirty or infected udders or from milkers' hands can result in mastitis.

Monitoring the quantity and quality of milk is done through herd testing, SCC and other quality tests. The amount of milk produced by a cow is usually estimated using herd testing, as many sheds do not have meters for measuring yield. In herd testing, an individual sample is taken from each cow, measured for volume and tested for milk composition (fat and protein) and SCC. Companies such as the LIC or CRV Ambreed (the second-largest breeding company in New Zealand) have technicians that come to the farm on set dates and set up milk sampling meters; the farmer then collects the milk samples and the technicians collect them from the farm and send them to the lab for analysis. The most common herd test regimen is four tests per season, although this is farm dependent. The amount of milk produced by the herd is measured every time the milk is collected from the vat, and reported back to the farmer through supplier-specific web pages (where farmers can only see their own farm) or mobile apps, but more commonly via a printed tanker docket.

The major focus for milk quality is mastitis, which is inflammation of the mammary gland (udder) caused by bacterial infection and one of the major health problems of dairy cows. It may be found in one or multiple quarters. Cows with SCCs over 150,000 cells/ml and heifers with SCCs over 120,000 cells/ml are considered to be showing signs of infection. A cow with subclinical mastitis shows no detectable signs, but an elevated SCC is detected in the milk through cow-side tests such as the rapid mastitis test (RMT) or a conductivity meter.

The RMT requires milk from each quarter to be put in a paddle and mixed with detergent — this breaks down cell membranes and turns the milk syrupy if the SCC is elevated. Sodium (Na) and chloride (Cl) ion concentrations change with an elevated SCC, and this is picked up by the conductivity test. Herd testing gives SCC results for individual cows. Subclinical cows are usually left to self-resolve, but can be treated with antibiotics depending on what organism is causing the infection.

Clinical mastitis is associated with physical signs such as behavioural changes (e.g. kicking the cups off), hot, swollen and hard teats, and clots and blood in the milk. It is usually detected when the filter sock (which covers the filter in the milk line) is examined at the end of milking and clots are detected. All the cows in the herd will then be 'stripped' (have milk expressed from each quarter for examination) prior to the next milking to find infected cows.

Once clinical cows are identified, they are milked out and treated with antibiotics. It is important to mark these cows and milk them separately so that their milk does not go into the vat. Antibiotics must be withheld from milk for certain periods after treatment has finished. There are large financial penalties if antibiotics are identified in the milk, and the tests are very sensitive; if milk containing antibiotics is put in the vat before the end of the withholding period, it can be detected.

Milk quality tests are carried out by the milk processor every time milk is picked up from the vat at the farm. The milk is tested for antibiotics, SCC, bacterial levels indicating dirty plant (incorrect cleaning of the milking machine) or dirty animals, and temperature. There can be financial penalties for failing these tests.

Pastures

Milkers (consuming 14–16 kg DM/head/day) and dry cows (consuming 9–12 kg DM/head/day) will be fed a mixture of pasture and supplements until pasture growth is sufficient for supplements to be stopped or, in some herds, reduced to 1–2 kg DM/head/day. Cows will have magnesium added to their diet to prevent metabolic diseases such as milk fever.

As pasture growth rates increase, cow feeding levels will be increased as they approach peak lactation and nitrogen fertiliser may be applied to help stimulate pasture growth rates. If there is heavy rainfall, cows will be removed from pasture and will stand off in areas such as races and shed yards for several hours a day to protect pastures from 'pugging' damage. Damaged pasture takes longer to recover and grow.

SPRING TO EARLY SUMMER (OCTOBER TO DECEMBER)

Mating

A cow does not produce milk until she has calved for the first time, at two years of age. To do this, she must have been mated and become pregnant nine months beforehand. Mating and pregnancy, and thus lactation, happen on a yearly basis on a New Zealand farm. Pregnancy lasts approximately 282 days. Every cow in the herd is expected to produce a calf every year, so the cows have approximately 83 days (365 – 282 = 83 days) after calving to recover from pregnancy and calving and to start cycling (moving into oestrus) ready to be mated and get pregnant again. If cows calve later than 282 days, this reduces the recovery time and increases the risk of the cow not getting pregnant (being 'empty') by the end of the mating period and thus being culled.

Heifers and cows have regular oestrus cycles of about 21 days. Behavioural changes indicate when they are ready to be mated. Farmers use aids to help identify cows that are in oestrus (on heat or bulling). These include tail paint, pressure detectors, scratchies (e.g. Estrotect) and activity meters. Pre-mating heats should be monitored so that the number of cows not cycling by the start of mating can be checked by a veterinarian to identify potential problems.

Cows can be mated using artificial insemination (AI), or naturally using bulls. For AI, semen (either fresh or frozen) is purchased from breeding companies such as the LIC that breed and manage the bulls and produce semen. Farmers identify and draft out cows that are in oestrus, and an AI technician from the breeding company visits the farm to inseminate the cows. Farmers who have trained appropriately or are AI technicians can artificially inseminate their own herds.

In most herds, mating lasts for 10–12 weeks. During the first 5–6 weeks cows will usually be bred using AI, after which bulls will be placed with the herd and will mate cows as they come into oestrus. The bulls used for mating dairy cows are usually Jersey bulls or beef bulls (Hereford or Angus). This makes it easy to identify the calves produced by the bulls, as most farmers do not retain replacement heifer calves from these matings. Jersey bulls are also commonly used for mating heifers, as they produce small calves which are easier to give birth to. Some farmers use only AI, as there may have been problems with running bulls due to their behaviour. Bulls may be aggressive and/or destroy fences. If bulls are purchased and brought onto a property, they need to have breeding soundness tests before mating begins. On some farms bulls may be retained for multiple seasons, whereas on others they will be sent to slaughter after mating has finished. Replacement heifer calves are usually selected from AI matings and early-calving cows.

The length of the mating period is important as it affects the calving spread for the following season. This determines days in milk (days lactating, DIM) and hence the productivity and profitability of the farm.

The InCalf programme is a national programme run by DairyNZ that aims to improve the reproductive performance of herds. Selection for milk production has led to reduced fertility in the modern dairy cow, but this is now slowly improving. The programme has targets that farmers can use as benchmarks for improving reproductive performance in their herds. It also helps them focus on areas that are of concern.

The main areas monitored are submission rate, conception rate, six-week in-calf rate and empty rate. The submission rate shows the number of cows that have been mated by AI over a set period of time; for example, a three-week submission rate of 90 per cent means that 90 per cent of a herd's cows should have been put up for mating in the first three weeks of mating. The six-week in-calf rate (the proportion of cows in calf by week six of mating) has a target of 78 per cent. However, given that the conception rate for AI is on average 53 per cent, this is higher than is achieved by many herds. The targets are set so that a compact calving spread results, and 87 per cent of the herd should calve by six weeks after the start of calving. The empty rate (number of non-pregnant cows) after 12 weeks of mating should be no more than 6 per cent.

Cows that have recently calved should resume cycling within six weeks of calving; those that have been identified as not cycling 4–5 weeks before mating starts need a veterinary check; they may have a uterine infection, which can be treated with antibiotics. If they are healthy but have not resumed cycling, hormone intervention can be used. In the past, cows that became pregnant late in the mating period were induced to calve through a course of hormone injections. However, due to animal welfare concerns induction has now been banned except when the welfare of the cow may be reduced (such as a bull having very large calves, causing dystocia or difficult births), as induced calves were born prematurely and either did not survive or were euthanased. This was also very stressful for both the cow and the farm staff.

Milking and calves

Milking continues, and cows should be at peak lactation and intake during this time. High-producing cows will require over 20 kg DM/cow/day. Calves will be weaned, and need to be grazing high-quality pasture. They need to be vaccinated at weaning and should be drenched regularly for gastrointestinal worms.

Pastures

Nitrogen fertiliser is applied to stimulate pasture growth. Turnip and maize crops are sown. When pasture growth exceeds what is required, paddocks are shut up to make silage and hay. Silage is cut forage that is stacked in a pit and undergoes a fermentation process that acts as a preservative. Many forages can be ensiled, such as pasture and lucerne, enabling surplus feed to be stored and used when feed supply is restricted, such as in summer and winter. Maize crops that are planted at this time are harvested for silage in March.

SUMMER TO AUTUMN (JANUARY TO APRIL)

Milking continues. Individual cow yields start to decrease. As reduced rainfall will usually limit pasture growth, cows will start to be dried off when yields are less than 5 litres/day. On some farms, rather than feed silage that is being saved for winter use, cows will be milked once a day to reduce feed demand. When cows are switched to OAD milking, increased monitoring of SCCs and the cows must occur to prevent an increase in mastitis.

Cows and 18-month-old heifers are pregnancy tested. If pasture supply is restricted, non-pregnant animals can be culled to reduce feed demand. Crops such as turnips will be fed. Recently there has been an increase in farmers using forage herbs such as plaintain and chicory, and legumes such as red clover and lucerne. These plants have a long tap root, enabling them to use soil water that is beyond the depth that ryegrass can reach. This results in much greater production of high-quality feed during dry summer months. Grass silage is also fed if it is available.

If red clover or lucerne is being fed, animals must be watched for bloat. Bloat can be prevented by transition feeding (feeding increasing amounts over several weeks until the rumen has adjusted) or adding a bloat oil to the water. Facial eczema (FE) or ryegrass staggers may occur with old pasture, and it is advisable to administer zinc (in water or by pill) to prevent FE.

When cows are dried off, some farmers will use dry cow therapy — a long-acting antibiotic which is administered through the teat into the udder to prevent mastitis developing over the dry period or when the cow calves in the spring. An alternative is to use a teat seal, which forms a physical barrier that stops bacteria from entering the udder. Farmers may use both of these methods. In some herds all cows are treated, while in others only selected cows will receive treatment. Teat-sealing heifers is recommended to prevent mastitis, which is common when heifers first calve.

Cows are culled for a variety of reasons, the most common being low milk production or

Dairy calves drinking milk on a calfeteria.

being 'empty'. On average, 22 per cent of cows are replaced with first-calving heifers for the start of each new season; older cows or those with bad temperaments will also be culled. Culling usually occurs strategically at certain times to help reduce feed demand.

Fertiliser is applied in autumn — mainly phosphorus (P), sulphur (S) and potassium (K). Minerals such as selenium can be applied with fertiliser. If it has rained, nitrogen fertiliser can be applied to increase pasture growth. The pasture response rate to nitrogen is lower in autumn than in spring, so growth is slower and less grass can be expected.

WINTER
(MAY TO JUNE)

Cows are dried off and fed to gain condition for calving. Thinner and younger cows should be preferentially fed to gain condition and grow. Nine-month-old heifers should be weighed to check that they are on target, and drenched.

Farm maintenance is conducted at this time. The jobs are repairing races and fences, having the milking plant tested, and carrying out routine maintenance and repairs as required.

Pastures are grazed intensively, silage (grass/maize) is fed and winter crops such as swedes and kale are grazed. To try to avoid excessive pasture damage (pugging) in wet conditions, some farms have sacrifice paddocks, stand-off pads or feed pads where cows are put for a portion of the day. The length of rotation through the different areas is set with the aim of having pasture cover of 2000–2400 kg DM/ha, sufficient to feed cows generously after calving. Pasture should be in good leafy condition and capable of rapid spring growth when temperatures rise.

This is a common time for farming families to take holidays, as it is the slowest time of the season. It is also the time when sharemilkers change farms. Contracts are set up so that changing between farms happens at the end of the season. 1 June is known as 'gypsy day', as this is when new sharemilking contracts start and so sees sharemilkers moving cows, machinery and families.

Jersey cow, body condition score 3.0.

Dairy operating structures

Traditional operating structures on New Zealand farms are described as owner-operator, sharemilker or contract milker. Owner-operators are farmers who own and operate their own farms, receiving all of the farm income. They comprise the largest group (67 per cent of all herds) (LIC & DairyNZ, 2015).

Sharemilking is a system of share farming, and has traditionally been a step to farm ownership. It is a contracted agreement between farm owners and farm workers, whereby the farm is operated on behalf of the farm owner for an agreed share of the farm profits (as opposed to a set wage). Common agreements are 50/50 or variable order (less than 50 per cent).

- Under a 50/50 agreement, the sharemilker owns the cows and any equipment other than the milking plant required to operate the farm. The sharemilker is responsible for milk and stock costs, general farm work and maintenance. The owner is responsible for maintaining the property. Fifty per cent of the milk income is received by the sharemilker and 50 per cent by the farm owner. The sharemilker receives most of the stock sales.
- Variable order is when the share-farming agreement is for a smaller percentage of the milk income and costs. The farm owner may still own the cows, and has a much greater involvement in the day-to-day running of the farm and in decision-making.

Contract milkers are employed to run the farm for a set price per kg MS produced.

Feeding

The feed eaten by cows is measured in kilograms of dry matter (kg DM), regardless of whether they are grazing pasture, forages or crops or are eating concentrates. Dry matter is defined as the solids left after all the water is taken out of the feed. The dry matter contains energy in the form of sugars and carbohydrate, which is measured in megajoules of metabolisable energy (MJ ME). The dry matter also contains protein and fibre, which are important components of feed. Together, these components define the nutrient content or quality of the feed.

As pasture ages, ME content decreases and fibre increases. This creates a low-energy high-bulk feed, which limits intake due to the physical constraints of the rumen (gut fill) and the increase in the time it takes to be digested. Therefore, consideration of both quantity and quality (Figure 1.3) is important in maximising feed value and production.

When pasture is growing fast (as in spring), the rotation length (time it takes to move through and graze all paddocks on the farm) is short (e.g. less than 21 days), so as to utilise pasture as effectively as possible. Depending on the stocking rate, the diet at this time may be all pasture or may be supplemented by a small amount of other feed. Cows are often break-fed (strip grazed) behind an electric fence during this time to maximise grass utilisation.

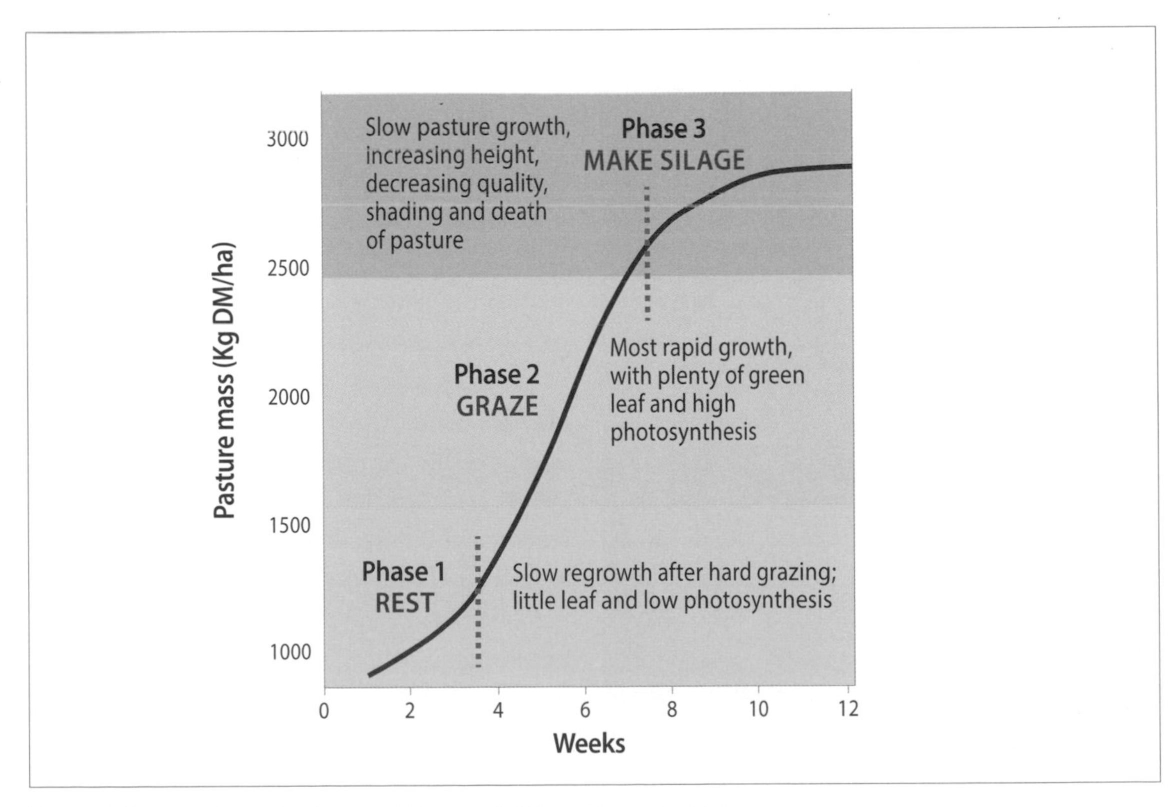

Figure 1.3 Reasons for grazing between 2500 and 1500 kg DM pasture mass.

A rising plate meter is used to measure the height of pasture.

Pasture growth rates slow in summer and autumn due to low rainfall, and crops such as turnips and fodder beet may be fed. Cows must be transitioned onto these crops to allow the rumen time to adjust to the new feed — the cows are allowed a small amount of crop in the beginning, which is increased over several weeks. Turnips are also fed, at a level of no more than 2–3 kg DM per cow per day to ensure that certain compounds in the turnips do not build up in the milk and cause milk taint. Milk taint gives the milk an undesirable taste and smell that can also transfer into any product made from the milk. Fodder beet also requires careful transitioning to avoid cows developing acidosis (acidic condition in the rumen which causes illness).

When pasture growth is very slow during winter, the rotation length is long (e.g. 100 days). This is so that pasture saved in the autumn may last as long as possible, until grass growth rates increase in the spring. At this time the cows' diet may contain a large proportion of feed other than pasture, such as grass silage, maize silage or crops such as kale, winter oats, fodder beet or swedes.

Farmers usually visually assess how much cow forage is available in a paddock. The estimated amount multiplied by the area of the paddock and divided by the number of cows to be fed will give an estimate of how much is being eaten. There are tools available to help with this assessment. For ryegrass/white clover pasture, a rising plate meter can be used. The plate of the meter measures the height of the pasture below and an equation is used to convert this to kg DM/ha. The RPM is only calibrated for this type of pasture; for other pastures or crops, a quadrat or a sward stick should be used. A quadrat is a metal square of a known area; it is laid on the

Jersey cow, body condition score 5+.

pasture and the forage within the square is harvested and oven-dried to remove the water. This gives the dry matter percentage (DM%) of the feed and, through calculation, the amount available. A sward stick can be used for forage herb crops — the height on these sticks has been calibrated to kg DM/ha by carrying out hundreds of quadrat cuts.

Farmers use pre-grazing and post-grazing masses as a proxy for what has been harvested from a particular paddock. This involves estimating how much feed is in the paddock before and after the cows have grazed it. It cannot be assumed that everything has been eaten, as there will be wastage from cows walking over pasture and defecating on it, which they will not then eat. This factor is called utilisation and varies between 60 and 85 per cent, depending on how dry or wet the paddock is. In spring, when the pre-grazing mass gets above 3000 kg DM/ha, these paddocks are often shut up for silage as pasture quality is beginning to decline and there will be high wastage from trampling. The post-grazing residual mass is recommended to be 1500–1600 kg DM/ha to encourage the growth of high-quality, leafy pasture.

Body condition score

Farmers use body condition score (BCS) to monitor the nutritional status of cows. Body condition is important not only for milk production but also for ensuring that a cow gets pregnant. The BCS is a visual assessment of a cow's body fat reserves. It measures the fat reserves at certain points of the body, such as the backbone, ribs, hip bones, rump, tip of the tail and thigh (Figure 1.4). The BCS can provide information on the energy balance and therefore the adequacy of feeding levels, along with health status, and indicates what the future feed requirements are if condition needs to be put back on a cow. It can also be used to provide a consistent way of assessing cow condition. DairyNZ provides advice, easy-to-use booklets and training courses (DairyNZ, 2016).

In New Zealand, the body condition of dairy cattle is assessed on a scale from 1 to 10, in increments of 0.5. If any animals in a herd score below 3, urgent action needs to be taken. Cows in this condition cannot be transported. A cow with a BCS greater than 6 is considered obese.

A change in BCS of 1 unit equates to 6.58 per cent of cow liveweight. For example:

- for a 425 kg Jersey cow, a change of 1 BCS unit is 28 kg liveweight
- for a 475 kg crossbred, a change of 1 BCS unit is 31 kg liveweight
- for a New Zealand Holstein Friesian, a change of 1 BCS unit is 33 kg liveweight

Cows should be routinely assessed for body condition. In spring, BCS can be used after calving to identify overly light cows that need preferential feeding. In autumn, it can be used for identifying cows with low body condition that should be dried off early so as to regain condition before calving. Cows need enough time to put condition back on before calving, as they don't put any condition on in the last month of pregnancy due to the demands of the fetus.

Body condition score targets are mainly used at calving and mating. At calving, it is recommended that mature cows have a BCS of 5.0, with heifers and rising three-year-olds having a score of 5.5. No more than 15 per cent of a herd should be below BCS 5.0, and no more than 15 per cent above BCS 5.5. A cow's BCS will drop after calving, as in early lactation feed intake is insufficient to meet the nutrient requirements of lactation and cows metabolise body tissue to meet this demand. It is recommended that the average decrease in BCS for a herd after calving is no greater than 1, so that at mating the BCS is 4.0 for cows and 4.5 for heifers and no more than 15 per cent of the herd is below BCS 4.0; all cows should also be gaining condition (DairyNZ, 2016).

The benefits for cows and heifers of meeting BCS targets is that they will produce more milk, become pregnant earlier in the breeding season, and be less likely to develop metabolic disorders such as milk fever (calcium deficiency which causes weakness). Thin cows are more likely to get mastitis and uterine infections, resume oestrus cycling later and not become pregnant. However, cows with too high a BCS can also have problems. Fat cows will eat less than thin cows after calving and will lose more body condition, which puts them at increased risk of metabolic disease, reduced milk production and failure to become pregnant.

Breeding and genetics

Dairy cow herds are often a mix of colour and sizes, due to the use of different breeds which have definite breed traits, and crossbreeding. The main breeds currently in New Zealand are Holstein Friesian (34.7 per cent), Jersey (10.4 per cent) and the KiwiCross (Holstein Friesian x Jersey; 45.6 per cent) (LIC & DairyNZ, 2015). The Holstein Friesian cows are the larger of the two parent breeds, are black and white and produce a greater volume of milk (Table 1.7). Jersey cows are smaller, brown cows with a higher percentage of fat and protein in their milk. As the figures show, the most dominant breed in the national herd is now the KiwiCross. This cow produces milk that is intermediate in volume and composition between the two parent breeds but also has good fertility (one of the benefits of crossbreeding). There are still small numbers of other breeds, such as Ayrshire, Guernsey and Brown Swiss, in the country.

Table 1.7 **Comparison of yield and composition of milk and liveweights between the three main breeds in New Zealand**

	Holstein Friesian (HF)	Jersey	KiwiCross (HF x Jersey)
Milk (litres per season)	4530	3266	4079
Fat (%)	4.46	5.69	4.97
Protein (%)	3.72	4.18	3.95
Fat yield (kg per season)	199.8	184.9	200.7
Protein yield (kg per season)	167.6	136.4	160.3
Liveweight of 4- to 8-year-olds (kg)	494	419	463
kg MS produced per kg liveweight	74	77	78

Source: adapted from LIC & DairyNZ (2015).
kg MS = kilogram of milksolids (fat + protein).

It must be acknowledged that regardless of breed, a high-producing cow is a high-producing cow. The choice of breed on a farm will be based on what each farmer believes best suits their system.

While the amount of milk produced by a cow is influenced by feeding, it also depends on age and genetic potential. A cow calves and starts to produce milk at two years of age. For the first two lactations, younger cows do not produce as much milk as older cows (approximately 75 per cent for the first lactation and 87 per cent for the second lactation). This is because they are still growing, in terms of both body size and the amount of mammary tissue needed to synthesise milk. Cows are fully grown at four years of age and most productive between four and seven years of age (Table 1.8).

As discussed previously, cow productivity is influenced not only by management but also by each cow's inherent abilities (i.e. its genes), which are inherited from her dam (mother) and her sire (father). To increase the rate of genetic gain and the ability to use bulls with high genetic merit to inseminate more cows, artificial insemination is used for around 70 per cent of the cows in New Zealand (LIC & DairyNZ, 2015). Cows of high genetic merit are capable of producing milk solids at low cost, as they are efficient converters of feed to milk and are also healthy, fertile, easy to milk and easy to manage.

Jersey BCS 3.0

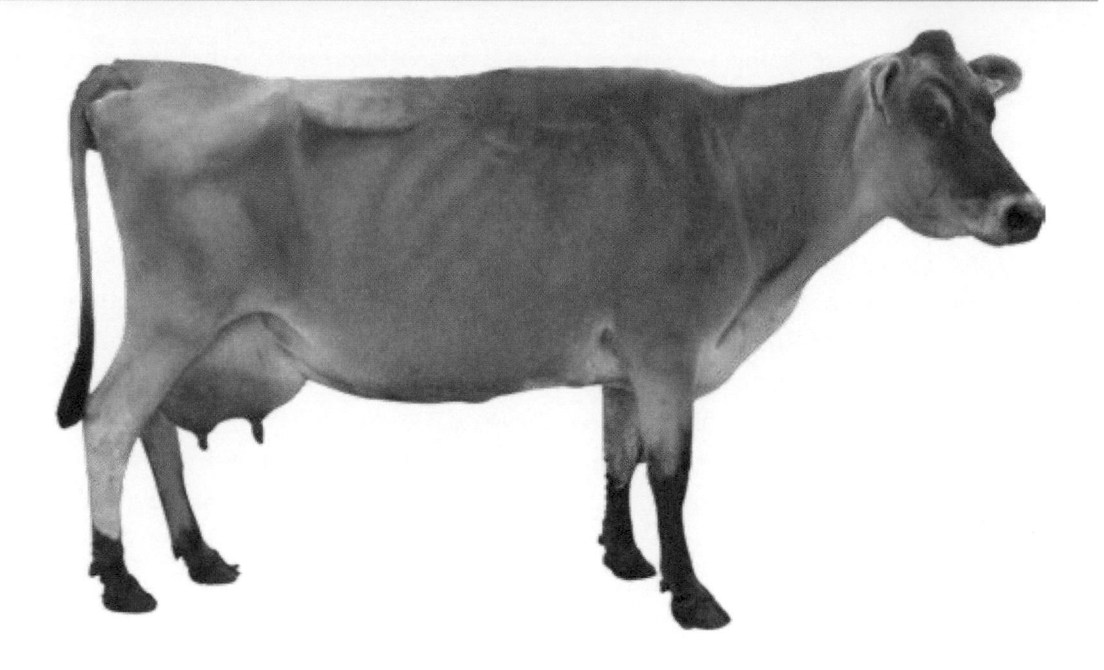

BACKBONE	Prominent ridge. "Roofing-iron" corrugations.	
LONG RIB	4 to 5 ribs easily seen.	
SHORT RIB	Prominent with edges sharp to the touch.	
HIP	Angular, sharp edges. Depressions on sides appearing.	
PINS	Tap-like appearance. Sharp edges.	
TAILHEAD	Deep "V" shape depression. Tailhead prominent, bumpy profile.	
RUMP	Deeply dished.	
THIGH	Indented. No visible fat. Muscle structure defined.	

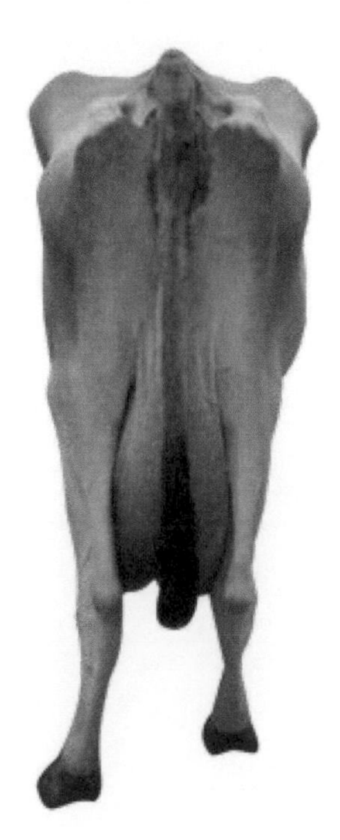

Figure 1.4 Body condition score. Source: DairyNZ, Body Condition Scoring Made Easy booklet.

Jersey BCS 5.0

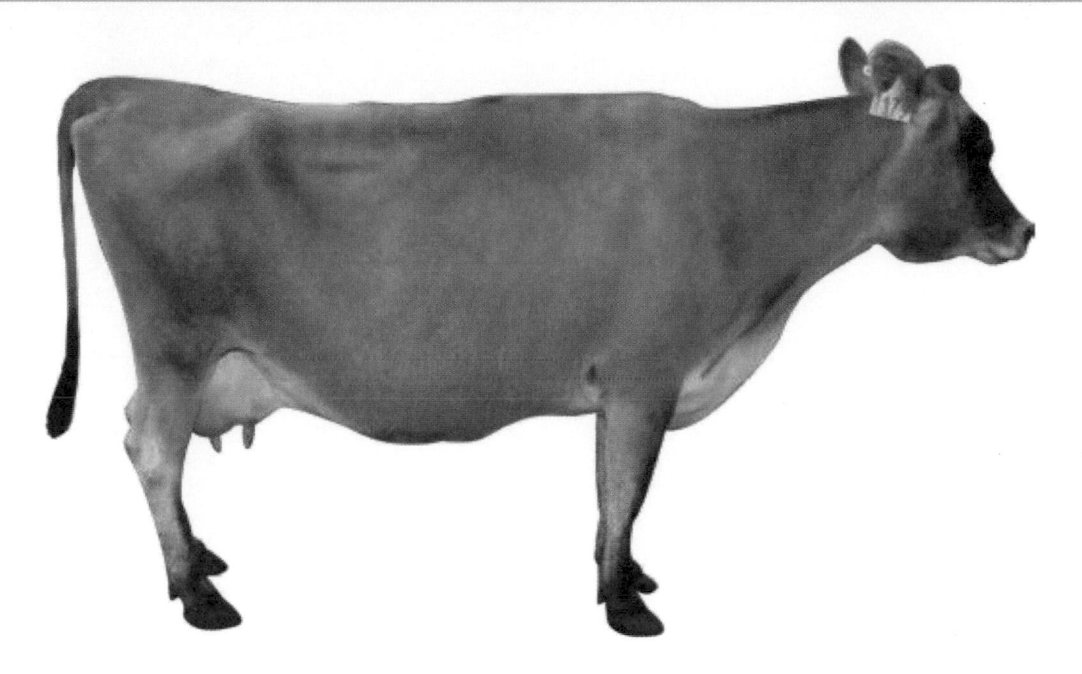

BACKBONE	Ridge easily visible but rounded and smooth.	
LONG RIB	Not visible but rounded to the touch.	
SHORT RIB	Rounded, individual ribs not visible but can be felt.	
HIP	Rounded. Curved in profile.	
PINS	Rounded.	
TAILHEAD	Tail rounded. Depression under tail filled. Even, no sharp edges.	
RUMP	Flat even cover.	
THIGH	Smooth and flat.	

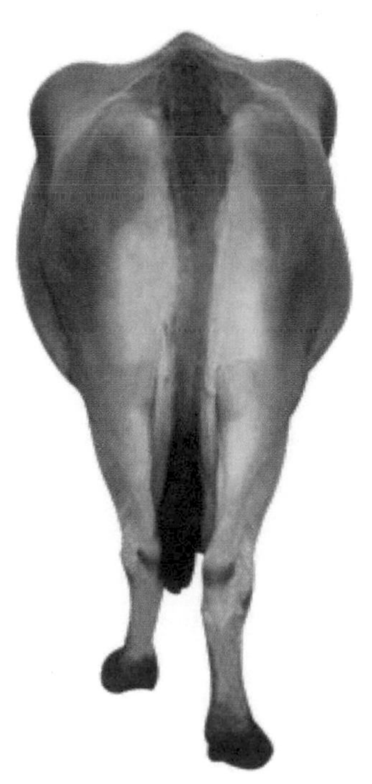

Table 1.8 **Effect of cow age on milk production**

Age (years)	Number	Milk (litres)	Milk fat (kg)	Milk protein (kg)	Liveweight (kg)*
2	297,811	3319	163.4	130.8	369
3	249,244	3970	194.3	157.3	423
4	204,903	4435	213.2	171.9	451
5	183,473	4491	221.0	176.7	471
6	150,810	4506	222.9	173.5	485
7–9	217,285	4360	215.4	169.0	465
10+	46,212	3898	177.0	148.2	479

Source: adapted from LIC & DairyNZ (2015).

*Weights have been calculated from a lower number of animals than for the milk variables.

In the New Zealand dairy system, a large proportion of farmers' income is derived from milk sales; therefore, yields and costs (rearing, reproduction, health, labour) need to be included in the evaluation of breeding value and genetic merit. The ranking of animals by genetic merit uses an index based on accumulated breeding values and their relative economic values. The national breeding objective of the New Zealand dairy industry is to select cows that are efficient converters of feed into profit. The index, known as breeding worth (BW), ranks both male and female animals based on their expected ability to breed profitable and efficient replacements, expressed as net income per 5 tonnes of dry matter eaten. It is calculated by combining certain breeding values with the appropriate economic values (New Zealand Animal Evaluation Ltd, 2015).

Before being used in the wider industry, bulls are first tested in a sire proving system, where a more accurate estimate of the genetic merit of a bull can be determined. The sire proving system allows the performance of a bull's daughters to be monitored over different environments and different farms. It also allows any genetic disorders to be identified before the bull has wider use in the industry.

Estimated breeding values (EBVs) are used to provide an estimate of a cow or bull's genetic merit for a particular trait — for example, will his or her daughters produce a greater yield of milk or have a greater liveweight? Relative economic values (EVs) are an estimate of the value of a 1 unit increase in the relevant trait to a New Zealand dairy farmer, assuming that the other traits remain unchanged.

Eight traits are now included in BW: fat, protein and milk yields, mature liveweight, fertility, somatic cell score, residual survival and BCS (NZAEL, 2016a). Breeding values for milk, fat, protein and somatic cell score are derived from herd test records. Fertility EBVs are derived from mating and calving records for cows on farm; liveweight and BCS EBVs are obtained from liveweight measurements taken during sire proving, and EBVs for residual survival are obtained from survival records after accounting for productive and reproductive causes for removal from the herd. Milk, protein and fat yields, and liveweight, are all expressed in kilograms; fertility is expressed as the percentage of a bull's daughters calving within the first 42 days of the herd's calving period; somatic cell score and BCS are expressed as numbers (scores); and residual survival is expressed in days (NZAEL, 2016a).

Economic values for fat, protein and milk volume are calculated using a five-year rolling average of the milk price. The liveweight EV includes the opportunity cost of feed, replacement heifer costs and cull cow/bobby calf values. The somatic cell score EV includes value lost through lowered survival, price penalties on milk supplied, and costs associated with clinical mastitis. The fertility EV includes value gained through increased survival and value gained through earlier calving dates and longer lactations. The residual survival EV includes value gained through rearing fewer replacement animals. The EV for BCS includes value gained through more days in milk or reduced feed costs to replace lost body condition (NZAEL, 2016b).

The EBVs are calculated using information collected on farm by farmers, e.g. herd testing, weighing data, BCS and 'traits other than production' scores. Cows that are routinely measured will have more accurate EBVs. When a heifer is born, her initial EBVs will be an average of those of her parents — hence the importance of correctly identifying the mother of a calf at birth. The accuracy of a bull's EBV will increase as more of his daughters are measured (progeny testing).

Each of the traits has a different heritability: milk volume 34 per cent, fat 30 per cent, protein 28 per cent, liveweight 35 per cent, somatic cell score 9 per cent, fertility 9 per cent, residual survival 6 per cent and BCS 25 per cent. The heritability of a trait provides an estimate of the variation that can be explained by genetics (NZAEL, 2016c).

Breeding worth and EBVs are expressed relative to a 'genetic base cow', which is updated every five years to reflect genetic progress. The current base cow is the average of a group of cows born in 2005 for which there are good records. An EBV of +10 kg protein indicates that a bull will transmit 5 kg more protein per lactation to his daughters than a bull with a protein EBV of 0; the +10 kg is halved because a bull, on average, passes on half of his EBV to his daughters.

Breeding worth is used both to rank bulls for breeding and to select cows from which to retain replacement calves. Farmers also have access to two other ranking tools to determine an animal's suitability for production: production worth (PW) and lactation worth (LW). Production worth estimates the ability of a cow to convert feed into profit over her lifetime compared with the average cow born in 2005 (the base cow); it is used for purchasing replacement cows and culling decisions. Lactation worth estimates the ability of a cow to convert feed into profit in the current season, and is used for culling decisions.

Genetic gain in dairy cows can be expressed as BW units per year. The current rate of genetic gain is around 10 units of BW per year, meaning that every five years the calves born have a potential to be $50 more profitable than the calves born five years earlier.

Animal health

Maintaining high health standards is crucial if dairy farms are to be productive; thus, dairy farmers spend a significant amount of time and money on prevention and treatment of disease. The three most important issues are fertility, mastitis and lameness. In calves, diarrhoea (or scours) is the most important disease. For all stages productivity is dependent on ensuring that nutritional deficiencies do not occur.

Fertility

For optimal milk production, dairy cows need to calve approximately once a year. This means that cattle need to become pregnant, on average, within 83 days of calving. Achieving this requires many changes to occur in the reproductive tract, particularly in the uterus. In most cases these changes take less than 60 days to complete, so there is more than enough time between calving and mating for restoration of the reproductive tract to take place.

However, if the process is delayed then cows may not become pregnant within the time available. There are two key reasons for failure to become pregnant. The first is infection of the uterus (endometritis), which is usually diagnosed either through the presence of discharge (pus) at the vulva or by using a special device to check for pus in the vagina, as cows with endometritis are not visibly sick. Infection of the uterus slows down the normal process of restoration of the reproductive tract, particularly in the ovaries and the uterus. Treatment of the infection returns the process of restoration to normal, so that cows can become pregnant within the normal breeding season. Two treatments are used for endometritis: a single dose of an antibiotic that is placed directly into the uterus; or a hormone, prostaglandin, which stimulates the ovary to start normal function that then results in the uterus clearing the infection. The disadvantage of prostaglandin treatment is that it requires the ovary to be in a specific state to work, whereas antibiotics are effective whatever state the ovary is in.

The second key fertility problem is the more important: failure of the ovary to start its normal cycle before the start of the breeding season. In these animals, ovulation does not occur so the animal fails to come into heat, is not inseminated and cannot become pregnant. Such cows are known as 'non-cyclers'. In many cases the failure is simply a consequence of the affected cows calving too close to the start of the breeding season (i.e. they have not had their 60 days), but in some cows the ovary has not started working properly even though the interval since calving has been sufficiently long.

In both cases the best short-term solution is to treat the cows with a combination of hormones over a nine-day period; this 'synchronisation programme' stimulates ovulation and means that the cows can be artificially inseminated early in the breeding season. Over the long term, alongside the use of hormones, the best solution is to work to improve fertility by optimising body condition, growing heifers well and making sure that feed quality in spring is ideal. Over time, cows will become pregnant and calve earlier and will be less likely to become non-cyclers. On the best-managed New Zealand farms, over 80 per cent of cows become pregnant in the first six weeks of the breeding season and 90 per cent of cows calve within the first six weeks of the calving season.

Mastitis

The most important infectious disease of dairy cattle in New Zealand is mastitis, or infection of the udder. Mastitis causes changes in the milk (clots or changes in colour) and, when severe, changes in the udder (reddening and/or swelling). Mastitis is an extremely expensive disease. Milk from cows with mastitis cannot be sold for processing, but this is not the only cost. There is also a treatment cost; plus, cows with mastitis produce less milk over the remaining lactation period, become pregnant less quickly, and lose more body condition. In addition, mastitis has a major welfare impact as it causes discomfort and distress. These factors mean that farmers spend a significant amount of time preventing, checking for and treating mastitis.

Almost all of the mastitis in dairy cows in New Zealand is caused by bacteria. Antibiotics are therefore the mainstay of treatment. Once mastitis is identified, the affected quarter is treated directly using a course of antibiotics placed directly into the udder via the teat canal. In most cases the response to treatment is rapid, with most cows receiving only three or four treatments over two days. Antibiotics are also used to treat infections when cows come to the end of their lactation; long-acting dry cow antibiotics are inserted into all four quarters of cows that are at risk of having infections at this time. Cows can also be given teat sealants at drying off. These have no antibiotic activity; they simply block the entry of bacteria into the udder via the teat. Their use can reduce the use of antibiotics at drying off. Teat sealants are also commonly used in pregnant heifers, with the product being inserted about one month before they calve for the first time.

Lameness

Lameness is the most important condition affecting the welfare of dairy cattle worldwide. In New Zealand, the level of lameness is very low by international standards, because almost all New Zealand dairy cattle are kept on pasture throughout the year. Nevertheless, approximately 20 per cent of cattle become lame at least once during the year. Most lameness is caused by damage to the horn-producing tissue, which then results in an inferior-quality horn that does not stand up to the normal day-to-day wear and tear on the hoof.

Under New Zealand conditions, the key risk factors for lameness are staff patience and approach (quiet, unhurried handling reduces the risk) and environment (e.g. walking tracks that are too small or badly maintained increase lameness risk). These two factors interact very strongly: for example, a badly designed milking shed which has poor cow flow is likely to lead to staff impatience and pushing to get cows through.

Early identification and treatment of lameness is central to reducing its impact. Watching cows as they walk to, or especially away, from milking is a good way of identifying lameness early on. Once identified, the affected foot needs to be lifted up and trimmed to remove damaged horn and reduce pressure on the affected claw. After treatment, lame cows are very commonly put into a 'lame cow mob' and kept in a paddock near the milking shed, and, often, milked only once a day. Combined with effective trimming, this helps the claw heal more quickly as the cow spends much less time standing.

Diarrhoea

A wide variety of pathogens can cause diarrhoea in calves, including bacteria, viruses and protozoa. Young calves under the age of one month are particularly prone to diseases as they have no immunity of their own and thus rely on getting this from colostrum (milk produced by the cow in the first 24 hours after calving). Prevention of diarrhoea thus focuses on ensuring that calves drink at least 10 per cent of their body weight in colostrum during their first 24 hours. Management of housing is also important — keeping calves in a warm, dry, well-ventilated but draught-free environment greatly reduces the risk of disease.

Irrespective of the cause, treatment of diarrhoea is directed towards reducing dehydration, replacing lost salts and stabilising the body's pH. In most cases this can be achieved by using specially designed electrolyte sachets that are dissolved in water. Antibiotics are not necessary in most cases.

Nutritional deficiencies

In a pasture-based system in particular, the most important nutritional deficiency is insufficient energy. This can be best monitored using simple techniques such as measuring pasture intake and monitoring BCS. However, insufficient intake of trace elements can also reduce productivity and cause disease, and these require more sophisticated techniques to diagnose.

The most important trace elements for dairy cows in New Zealand are selenium and copper, with cobalt and iodine being of importance in some herds. Deficiencies of these trace elements can cause diseases such as white muscle disease (selenium deficiency) or bush sickness (cobalt deficiency). Before these deficiency diseases develop, however, affected cattle become less productive (growth rates slow and milk production falls). Effective management of trace elements therefore relies on identifying animals that are at risk of deficiency (but are not yet deficient), rather than on diagnosing deficiencies. Routine testing of normal animals is thus required so that supplementation can be started (or stopped) before problems occur. This may involve blood sampling, or, especially for copper, liver sampling.

Environmental considerations

As discussed in previous sections, dairy farming in New Zealand is based on a predominantly grazed pasture system. However, concern is increasing regarding the environmental impacts of our dairy industry and its effect on surface and groundwater quality, something that is putting the 'clean green' image of the industry's export products under threat.

The main causes of dairying's adverse impact on water quality are nitrogen leaching and phosphorus run-off. Faecal-bacteria-contaminated run-off can also reduce the range of water uses; the source of these contaminants is mostly concentrated urine and dung patches that cows deposit when grazing (Ledgard & Menneer, 2005; McDowell et al., 2008).

Nitrogen leaching

The amount of nitrogen cycling in a dairy system is dependent on several factors, with pasture eaten by the cow being a significant component of the nitrogen cycle. Approximately 20 per cent of the nitrogen consumed is partitioned to meat and milk production, with the remaining 80 per cent being excreted as dung and urine (During, 1972; Haynes & Williams, 1993). For cows eating pasture, approximately 60 per cent of the nitrogen is excreted in urine, with 20 per cent in dung (Haynes & Williams, 1993; Ledgard & Steele, 1992); urine is therefore the main source of the nitrate leached from dairy farms. The concentrations of nutrients in dung and urine patches are very high. While nutrients are used for pasture growth, in urine patches an annual pasture production of up to 20 tonnes/ha would fail to use all of the available nitrogen prior to the winter drainage season, leading to problems with leaching.

Take, for example, 100 dairy cows grazing 1 hectare for a full day, ingesting 1200 kg DM of pasture and 350 kg DM of maize silage (Figure 1.5). On average, the pasture will consist of 2.5 per cent nitrogen and the maize silage 1 per cent nitrogen (Care & Hedley, 2008). The cows will therefore ingest 33.5 kg nitrogen per hectare per day. Of this, 5.7 kg will go towards milk production and 20 kg will be excreted as urine. A small amount of nitrogen (1.5 kg) will be volatilised from urine patches as ammonia (NH_3).

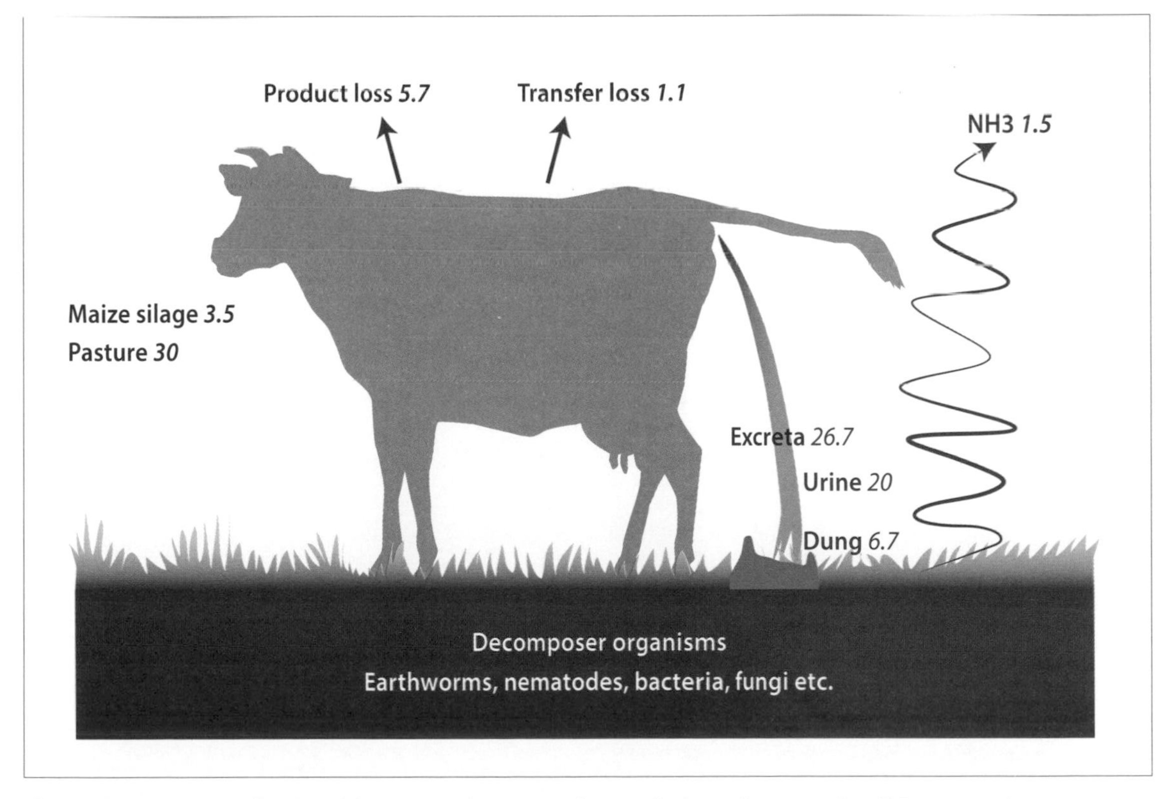

Figure 1.5 Nitrogen transfers in a dairy-cow grazing system, for one day's grazing, assuming 100 cows per hectare ingesting 1200 kg DM of pasture @ 2.5 per cent nitrogen and 350 kg DM of maize silage @ 1 per cent nitrogen. Numbers in italics denote kg of nitrogen. Source: adapted from Care & Hedley (2008).

Phosphorus run-off

Phosphorus is an important nutrient for all organisms within pastoral systems, and is fundamental in both tissue and skeletal development in plants and animals. It is also important for maintaining the rumen environment of grazing animals. For these reasons, New Zealand dairy pastures receive regular applications of phosphorus fertilisers to maintain optimum levels of plant-available phosphorus in soils (Roberts & Morton, 1999).

When cows graze, the deposition of dung returns phosphorus to the soil. If surface run-off occurs soon after grazing by dairy cattle, the amount of phosphorus lost in this run-off is greater than if the pasture had not been grazed recently (McDowell et al., 2006; Smith & Monaghan, 2003; Smith et al., 2001).

Faecal-bacteria-contaminated run-off

Cows are carriers of microorganisms that can be transferred to humans and cause health problems. Microorganisms such as campylobacter, giardia, cryptosporidium and E. coli are all sourced from faecal matter and can enter waterways through drainage or run-off (Collins, 2004).

Concentrations of microorganisms are generally higher in surface run-off from recently grazed pastures (Wilcock et al., 1999), due to dung being transported in run-off. The weather following grazing affects the amount of bacteria being lost through surface run-off. E. coli is often used as a faecal indicator organism (FIO) — if the concentrations of E. coli in surface run-off are high, this indicates that concentrations of other faecal organisms could also be high (Toranzos & McFeters, 1997). The entry of such organisms into waterways can pose serious health risks, with levels of 550 E. coli per 100 mL of water sampled from recreational waterways being deemed the upper limit for public safety.

The future

Considerable work has been done by regional councils and farmers through environmental programmes to fence waterways and restrictions in the amount of fertiliser usage allowed, to reduce nutrient leaching and run-off. However, the large flows of nitrogen, phosphorus and microorganisms via dairy cow urine and dung indicates that control measures to reduce the impacts of these contaminants on waterways should focus on the grazing area and time available for grazing in sensitive areas.

Animal welfare

The welfare issues of dairy cattle include painful conditions like lameness, mastitis and dystocia (Stafford, 2013). Malnutrition due to food shortages or deficiencies in specific trace elements is also important, as are a number of diseases (such as facial eczema) that can affect large numbers of cattle if not prevented. The management of cull cows after drying off and during transport for slaughter is sometimes inadequate, as is the management of unwanted calves that may be killed on farm or sold for slaughter for human consumption or pet food. The lack of shelter or shade during inclement weather is a significant issue that needs to be considered in all farming systems.

A code of welfare for dairy cattle (Anon, 2010) lists the minimum standards for dairy cattle. The amendments to the Animal Welfare Act 1999 that were introduced in 2015 list animal welfare transgressions that are fineable offences, including tail-docking cows or disbudding (removing horn buds) calves without pain relief.

Conclusion

To run a profitable and sustainable business, dairy farmers must balance feed demand and feed supply. To achieve this balance, events must occur on a seasonal basis on the farm and a variety of systems are used to do this.

The challenges faced by farmers include running a high-producing farm while meeting environmental standards. In addition, there is increasing scrutiny from the public about how their food is produced in terms of animal welfare. Another challenge is selling our products — the products that milk is made into are considered commodity or bulk products, which are easy to sell in large amounts. The development of new products and markets that take advantage of how and where our products are produced, and hence yield greater financial returns, is therefore essential.

The New Zealand dairy industry, its farmers and its farms are constantly evolving and will continue to meet these challenges with new solutions.

References

Anon. (2010). Code of welfare for dairy cows. Wellington: Ministry for Primary Industries, Government of New Zealand.

Care, D., & Hedley, M. (2008). Housing New Zealand's dairy cows. In I.M. Brookes (Ed.), *Dairy3* (pp. 167–77). Rotorua: Dairy3 Organising Committee, Rotorua.

Collins, R. (2004). Fecal contamination of pastoral wetlands. *Journal of Environmental Quality*, 33, 1912–18.

DairyNZ. (2016). Body condition scoring. Accessed at www.dairynz.co.nz/animal/herd-management/body-condition-scoring.

DairyNZ. (2010). Facts and figures for New Zealand dairy farmers. Accessed at www.dairynz.co.nz/publications/dairy-industry/facts-and-figures

During, C. (1972). Fertilisers and soils in New Zealand farming. *New Zealand Department of Agriculture Bulletin* No. 409.

Haynes, R.J., & Williams, P.H. (1993). Nutrient cycling and soil fertility in the grazed pasture ecosystem. *Advances in Agronomy*, 49, 119–99.

Holmes, C.M., Brookes, I.M., Garrick, D.J., MacKenzie, D.D.S, Parkinson, T.J., & Wilson, G.F. (2007). Milk production from pasture: principles and practices. Palmerston North: Massey University.

Ledgard, S.F., & Menneer, J.C. (2005). Nitrate leaching in grazing systems and management strategies to reduce losses. In L.D. Currie & J.A. Hanly (Eds), *Developments in fertiliser application technologies and nutrient management* (pp. 79–92). Palmerston North: Fertiliser and Lime Research Centre, Massey University.

Ledgard, S.F., & Steele, K.W. (1992). Biological nitrogen fixation in mixed legume/grass pastures. *Plant and Soil*, 141, 137–53.

Livestock Improvement Corporation (LIC), & DairyNZ. (2015). *New Zealand dairy statistics 2014–15*. Hamilton: LIC and DairyNZ.

McDowell, R.W., Houlbrooke, D.J., Muirhead, R.W., Muller, K., Shepherd, M., & Cuttle, S.P. (2008). Grazed pastures and surface water quality. New York, NY: Nova Science Publishers.

McDowell, R.W., Muirhead, R.W., & Monaghan, R.M. (2006). Nutrient, sediment, and bacterial losses in overland flow from pasture and cropping soils following cattle dung deposition. *Communications in Soil Science and Plant Analysis*, 37, 93–108.

McNaughton, L.R., & Lopdell, T.J. (2012). Are dairy heifers achieving liveweights targets? *Proceedings of the New Zealand Society of Animal Production*, 72, 120–22.

Mikkelsen, J.B. (2012). Factors that influence productivity and profitability of a once-a-day milking system (Honours thesis). Palmerston North: Massey University.

Ministry for Primary Industries. (2016). Accessed at www.mpi.govt.nz/exporting/food/dairy

New Zealand Animal Evaluation Ltd (NZAEL). (2015). Economic value update 2015. Accessed at www.dairynz.co.nz/animal/animal-evaluation/interpreting-the-info/economic-values/.

New Zealand Animal Evaluation Ltd (NZAEL). (2016a). All about BW. Accessed at www.dairynz.co.nz/animal/animal-evaluation/interpreting-the-info/all-about-bw/.

New Zealand Animal Evaluation Ltd (NZAEL). (2016b). Economic values. Accessed at www.dairynz.co.nz/animal/animal-evaluation/interpreting-the-info/economic-values/.

New Zealand Animal Evaluation Ltd (NZAEL). (2016c). Breeding values. Accessed at www.dairynz.co.nz/animal/animal-evaluation/interpreting-the-info/breeding-values/.

New Zealand Specialist Cheesemakers Association. (2016). Accessed at www.nzsca.org.nz.

Rattray, P.V., Brookes, I.M., & Nicol, A.M. (2007). Pasture and supplements for grazing animals. Occasional Publication no. 14. Cambridge, New Zealand: New Zealand Society of Animal Production.

Roberts, A.H.C., & Morton, J.D. (1999). Fertiliser use on New Zealand dairy farms. Auckland: New Zealand Fertiliser Manufacturers' Association.

Smith, K.A., Jackson, D.R., & Withers, P.J.A. (2001). Nutrient losses by surface run-off following the application of organic manures to arable land, 2. Phosphorus. *Environmental Pollution*, 112, 53–60.

Smith, L.C., & Monaghan, R.M. (2003). Nitrogen and phosphorus losses in overland flow from a cattle-grazed pasture in Southland. *New Zealand Journal of Agricultural Research*, 46, 225–37.

Stafford, K.J. (2013). Animal welfare in New Zealand. Occasional Publication no. 16. Cambridge, New Zealand: New Zealand Society of Animal Production.

Toranzos, G.A., & McFeters, G.A. (1997). Detection of indicator microorganisms in environmental freshwater and drinking waters. In C.J. Hurst (Ed.), *Manual of environmental microbiology* (pp. 184–94). Washington DC: ASM Press.

Wilcock, R.J., Nagels, J.W., Rodda, H.J.E., O'Connor, M.B., Thorrold, B.S., & Barnett, J.W. (1999). Water quality of a lowland stream in a New Zealand dairy farming catchment. *New Zealand Journal of Marine and Freshwater Research*, 33, 683–96.

Chapter 2

Beef Cattle Production

Steve Morris

Chapter 2

Beef Cattle Production

Steve Morris
Institute of Veterinary, Animal and Biomedical Sciences
Massey University, Palmerston North

Introduction

New Zealand has two kinds of cattle: dairy cattle and beef cattle. Dairy cows produce milk, and their calves are used as replacement heifers or are used to produce veal or beef. The calves from cows of the beef breeds are reared as replacement heifers or are grown to be killed as prime beef. In reality, however, virtually all cattle are eventually killed and eaten.

There are two common breeds of beef cattle — Angus and Hereford — and many other less common breeds. Beef for mince, often called manufacturing beef, comes from dairy bulls, especially Holstein Friesian bulls, and old dairy cows; higher-priced cuts of beef come from prime steers and heifers. Some dairy calves, known as bobby calves, are killed for veal.

The beef cattle population is basically a herd of beef breeding cows producing calves. The male (bull) calves are castrated and raised as steers for slaughter, either on the farm of origin or on finishing farms that are usually located on more productive land. Female (heifer) calves are either reared to replace the cows culled from the breeding herd, or raised for slaughter. This is the traditional beef cattle management system; other systems that rely on cattle from the dairy herd have also been developed. In one such system, bull calves (usually Holstein Friesian) are

Angus bulls. **Previous:** A typical hill-country sheep and beef farm.

purchased from dairy herds and raised as bulls for slaughter. Bull calves born to dairy cows mated with beef-breed bulls (crossbred calves) may be castrated and raised for slaughter. Crossbred heifer calves may be raised for beef or as beef breeding cows. The advantages for beef cattle farmers of using dairy-sourced cattle for beef production are that there is no capital overhead tied up in a beef breeding herd, and — as there is no breeding herd to feed — more grass is available for animals being raised for slaughter.

Beef cattle and sheep are often farmed together, and complement one another, especially on hill-country farms. Beef cattle may also be found on farms where most of the income comes from some other enterprise, usually sheep. Beef cattle graze with sheep and are used to maintain good-quality pasture for the sheep flock. In spring, however, the need to increase grass mass on areas grazed by ewes and lambs may delay the development of the high pasture masses and pasture allowances needed for high cattle liveweight gains.

Farmers alter their mix of sheep and cattle to suit the current economic conditions and preferences. During the 1970s the number of sheep increased and cattle numbers declined, driven by market prices that favoured sheep. The increase in beef cattle numbers seen since 1983 initially stabilised at 5 million head, but numbers have now declined to 3.6 million. The fluctuations in the number of animals killed for beef are caused by changes in the number of dairy calves reared for beef production.

The New Zealand beef industry

Around 13 per cent of total global beef production is traded internationally. New Zealand produces only 0.9 per cent of global beef production but accounts for about 6 per cent of the internationally traded volume. The beef industry plays a significant role in the New Zealand primary sector, with beef and veal exports worth about NZ$3 billion a year. New Zealand beef cattle are fed pasture, sometimes supplemented in winter with silage or haylage if a farm has suitable flat land for making these feeds. In many other countries, cattle are finished on feedlots and fed grain-based and crop-based diets.

India has the world's largest national cattle herd, with over 301 million head. Brazil and China have 213 million and 103 million respectively, while the United States is fourth with 88 million. In New Zealand the total cattle population is about 10 million, comprising 3.6 million beef and 6.4 million dairy cattle; it ranks as the twelfth-largest cattle herd by country. Australia ranks seventh in the world, with 27.6 million cattle. In New Zealand the annual per capita consumption of beef meat on a carcass weight equivalent basis is 24 kg; in Argentina it is 63 kg, Australia 32 kg, the United States 35 kg and China 4.6 kg.

An important factor that affects a country's position in the world beef trade is the health status of its cattle. New Zealand is free of the most serious cattle diseases — foot and mouth disease (FMD) and bovine spongiform encephalopathy (BSE) — and this allows it to export fresh chilled and frozen beef to a large number of markets. However, bovine tuberculosis (bTB) poses a threat to our beef exports. New Zealand has an active programme in place to control bTB and to ensure that it has a very low incidence. Our international cattle health status was bolstered in 2012 with the introduction of a compulsory individual-animal electronic identification (EID) programme. Compulsory EID allows tracking of cattle, which enables high-level disease surveillance and greatly enhances our biosecurity integrity and the confidence of our international customers.

The international beef trade comprises several sub-markets. Countries without a feedlot industry, like New Zealand, can supply grass-fed beef — nearly all of New Zealand's beef exports are indeed directed to markets for grass-fed beef, including the US and Asian markets. Australian grass-fed beef is also exported to the United States and Asia. The EU produces limited amounts of grass-fed beef, and exports it to Eastern Europe, Russia and Africa. South American countries have long been prohibited from sending fresh chilled and frozen beef to the high-value Asian and North American markets because of their FMD status. South America's major market has historically been the EU, which has allowed imports of such product from FMD countries subject to specific standards.

Two important features of the New Zealand beef industry are that about 95 per cent of feed is grazed pasture and that the dairy industry contributes a significant proportion of beef production. There is little scope for increased pastoral livestock numbers, as land development has largely been completed in New Zealand. Therefore, the relative areas of land used for beef, sheep or dairy livestock are determined largely by potential production and profit. The recent upsurge in dairy cow numbers has resulted in an

Table 2.1 **Beef and veal tonnes exported; shipped meat weight for the year ended 30 September 2014**

World region	Shipped meat weight (tonnes)	Percentage of total
North America	204,872	52%
North Asia	113,239	29%
South Asia	32,915	8%
Middle East	13,954	4%
European Union	11,524	3%
Pacific	10,198	3%
Other	4031	1%
Total	390,733	100%

Source: Beef & Lamb New Zealand Economic Service.

increased tonnage of cow beef, with declining steer, heifer and bull beef. Cow beef increased from 21 per cent of total beef production in 2000 to over 40 per cent in 2014.

North America is the dominant export market for beef, accounting for 52 per cent of beef exports by volume; North Asia, mainly China but also Japan, South Korea and Taiwan, accounts for 29 per cent of exports (Table 2.1).

Although beef exports are dominated by the North American market, there have been increased efforts in recent years to diversify to other markets in North and South Asia (i.e. China, South Korea, Taiwan, Japan, Hong Kong, Singapore, Indonesia and Malaysia). While this diversification will presumably continue, these new markets require prime beef as opposed to the traditional processing or ingredient beef required in North America. Nevertheless, beef exporting remains a relatively straightforward commodity business aimed at supplying frozen, boneless lean beef of consistent specifications to processors.

Approximately 80 per cent of all beef produced in New Zealand is exported, leaving 20 per cent (approximately 400,000 head per annum) for domestic consumption. Because the export market is the dominant market, the value of cattle is largely a reflection of what the licensed exporters are willing to pay. However, there is also a difference in carcass weights desired by the two markets, the local trade preferring carcasses weighing 220–240 kg which is 70–90 kg lighter than the average export steer or heifer carcass.

The major participants in the beef industry are farmers, meat processors, licensed exporters and transporters. The domestic and the export trade are generally separated, although some meat from licensed meat export plants does end up being used for domestic consumption. However, meat processed at domestic abattoirs cannot be sold for export.

There is a tendency to link processing and marketing together, but this is a misconception. Processing is preparing the product to the point where it is ready to be sold. Marketing occurs when the product is exported and sold. A meat company may be a processor of meat and licensed to export and sell meat in international markets. In New Zealand, most of the larger meat companies (AFFCO, Alliance, ANZCO and Silver Fern Farms) are processors and exporters.

A brief history of beef production

Beef cattle production in New Zealand has a long and eventful history. The Reverend Samuel Marsden (1765–1838) is credited with importing the first cattle into New Zealand. These cattle were kept primarily for milk production rather than for meat. During the early European settlement of New Zealand, cattle numbers increased slowly despite numerous importations. Concentrations of cattle were found in provinces like Otago, with its relatively large number of people attracted by the discovery of gold in the 1860s.

By 1861, cattle numbers in the South Island had reached 97,000 head, 1000 more than the North Island total. Ten years later, Otago and Southland had 143,000 head of cattle, while the province of Auckland had the next highest total of 80,000 head. With the advent of refrigerated shipping in 1882, the Auckland province soon became the largest producer of butter, cheese and beef from a cattle population that by 1886 had reached 205,000 head.

Thus, during its infancy the beef industry was an appendage of the dairy and transport industries. Until the mid-1950s, beef cattle production was regarded by farmers as a necessary adjunct to sheep finishing — cattle were kept to maintain pastures in a suitable condition for sheep.

Three major events are cited as important contributors to bringing about technological change in the production, processing and marketing of beef. The first was the successful development and introduction of refrigeration; the second was the rise of the chilled beef trade and its subsequent decline and eventual abandonment; and the third was the rapid development of the US market as the largest and most profitable outlet for New Zealand beef (Campbell, 1970).

The development of the chilled beef trade resulted in a swing to the Angus breed and the production of a chiller-type beast — small in size, compact in conformation and over-finished (over-fat) by present-day standards. The post-World War II cost of processing and shipping chilled bone-in quarters of beef, and the lack of a satisfactory premium, contributed to the decline and eventual end of this trade.

The rise of the more lucrative US market for New Zealand beef also helped to bring about the demise of the chilled beef industry. The opening up of this market in the early 1960s had a far-reaching impact on New Zealand. The prices received for beef on the US market, and the fact that the government of that country controls imports (via a quota allowance), have collectively given a measure of price stability to farmers for their cattle. These two factors were key in encouraging a rapid increase in cattle numbers.

The impact of this development on the farm has been dramatic. Beef cattle enterprises became profitable in their own right, and old ideas of what constituted a 'quality' beast were discarded. This led to a change in the type of animal bred for beef production and also to the rise of the dairy-beef cattle sector. This latter move has increased the supply of cattle for slaughter and has provided the exporter with a greater volume of lean beef of the quality demanded by the US market.

The changes in practices on New Zealand beef farms that have helped lead to these gains include increased investment in fertiliser, improved pasture production and management, implementation of pregnancy diagnosis, body condition scoring, use of terminal sires (bulls used to produce calves reared only for meat and not as replacement heifers), use of crossbreeding (including development of composite breeds), genetic selection for improved production, improved whole-herd health plans, and once-bred heifers (heifers that have calved and are then processed for prime beef).

Unlike many countries, the New Zealand beef industry has not used reproductive technologies such as artificial insemination (AI) and embryo transfer (ET) widely, due to the extensive nature of beef cattle farming in New Zealand and the logistics and cost of these technologies (Morris & Kenyon, 2014).

The processing and marketing of New Zealand beef has changed dramatically since 1985. There has been a major shift to specialist, one-chain beef processing plants, with an associated improvement in efficiency. Hot-boning plants have become fashionable (beef is traditionally boned after 24 hours of chilling), and these plants are considered world leaders in processing meat. Some operate two shifts per day, and primarily slaughter and process Holstein Friesian bulls for the US processing beef trade.

The national beef breeding herd has shrunk since the 1970s, with the herd having lost more than 900,000 cows (Table 2.2). The dairy industry now contributes significantly to beef production, with an estimated 35 per cent of the calves entering the beef industry each year having been born on dairy farms. It is also estimated that around 750,000 dairy cows constitute a major contributor to the 950,000 cows processed for beef each year.

During the spring of the 2015 season, about 450,000 dairy calves were moved onto beef production farms. This is about 33 per cent of the total number of calves entering the beef cattle herd. In addition, about 2 million bobby calves were slaughtered in 2015. If more of these calves were raised rather than slaughtered, national beef production could be increased significantly. Low beef prices increase the bobby calf kill, while high prices tend to increase the proportion of dairy-bred calves raised for beef production.

The higher percentage of beef cattle derived from the dairy herd (Table 2.3) has resulted in the ratio of beef breeding cows and heifers

Table 2.2 **Composition of the New Zealand beef herd (in '000s at 30 June)**

	Year		
	1973	**1993**	**2015**
Total beef herd	5343	4676	3576
Breeding cows	1907	1419	996
Other cattle	3436	3257	2580
Breeding cows as % of total	36	30	28
Other cattle as % of total	64	70	72

Source: Beef & Lamb New Zealand Economic Service.

Table 2.3 **Importance of dairy herd input into the national beef herd (in '000s at June 2015)**

	Number	Percentage
Number of beef cows	996	
Calving percentage (%)		85
Number of calves born to beef cows	800	67
Number of calves from dairy cows	450	33
Total calf input	1250	100

Source: Beef & Lamb New Zealand Economic Service.

Table 2.4 **New Zealand's beef and veal production 1990–2015**

Year	Production ('000 tonnes)	Slaughter ('000 head)	
		Calves	Cattle
1990	478	725	1945
1994	578	808	2045
1998	632	1339	2478
2002	553	1291	2181
2006	619	1388	2337
2010	609	1552	2376
2015	633	2128	2403

Source: Beef & Lamb New Zealand Economic Service.

Notes: (a) Year ending September 30; (b) Production is bone-in basis; and (c) Available for export is less than total, as some is withdrawn from export for local market.

Table 2.5 **Number of head of beef and veal processed for export**

Type	Year		
	1993	2003	2015
Steer	557,639	450,400	558,000
Heifer	199,639	229,000	398,000
Cow	587,291	833,900	1,187,000
Bull	610,037	652,900	488,000
Vealer	292	–	–
Bobby calf	812,802	1,550,000	2,119,000

Source: Beef & Lamb New Zealand Economic Service.

Table 2.6 **Trend in average export beef carcass weights (kg per head)**

Year	Steer	Heifer	Cow	Bull
1979–80	277.2	210.7	185.1	252.4
1989–90	291.8	217.7	182.1	277.1
1999–00	311.6	227.9	196.7	301.5
2005–06	323.6	243.4	204.4	313.6
2011–12	317.0	245.0	205.0	309.0
2014–15	302.0	234.0	197.0	298.0

Source: Beef & Lamb New Zealand Economic Service.

Hereford cows and calves on hill country.

in the national herd declining from 36 per cent in 1972–73 to 28 per cent in 2014–15, with a resultant increase in 'trading' or finishing stock.

For the year ending 30 September 2015, New Zealand produced 633,000 tonnes of beef. The increase in beef and veal production over the period 1990 to 2014 is shown in Table 2.4. Annual beef production over this decade has fluctuated between 478,000 and 633,000 tonnes (bone-in basis). Between 80,000 and 120,000 tonnes is consumed in the domestic market, while the remainder is destined for export. Of the adult cattle that were slaughtered for export in the year ended 30 September 2015, some 20 per cent were steers, 17 per cent heifers, 44 per cent cows and 18 per cent bulls (Table 2.5). An estimated 400,000 cattle (around 20 per cent of the adult kill) were slaughtered for the domestic market.

Carcass weights for steers, heifers, cows and bulls have all increased considerably over the past 20 years (Table 2.6). Average age at slaughter is not available, but it is estimated that 50 per cent of steers are slaughtered between 18 and 24 months of age, with the remaining 50 per cent often carried through on farm for a second winter and ranging from 27–34 months of age at slaughter. Assuming an average birth weight of 35 kg and a dressing-out percentage (carcass weight as a percentage of total liveweight) of 54 for a 300 kg carcass, liveweight gains averaging 1.0 kg/day are achieved on properties where cattle are slaughtered at around 300 kg carcass weight at 18–24 months of age. Steers slaughtered at 27–34 months of age will achieve lifetime liveweight gains of 0.5–0.6 kg/day. Heifers are usually slaughtered at 18–24 months

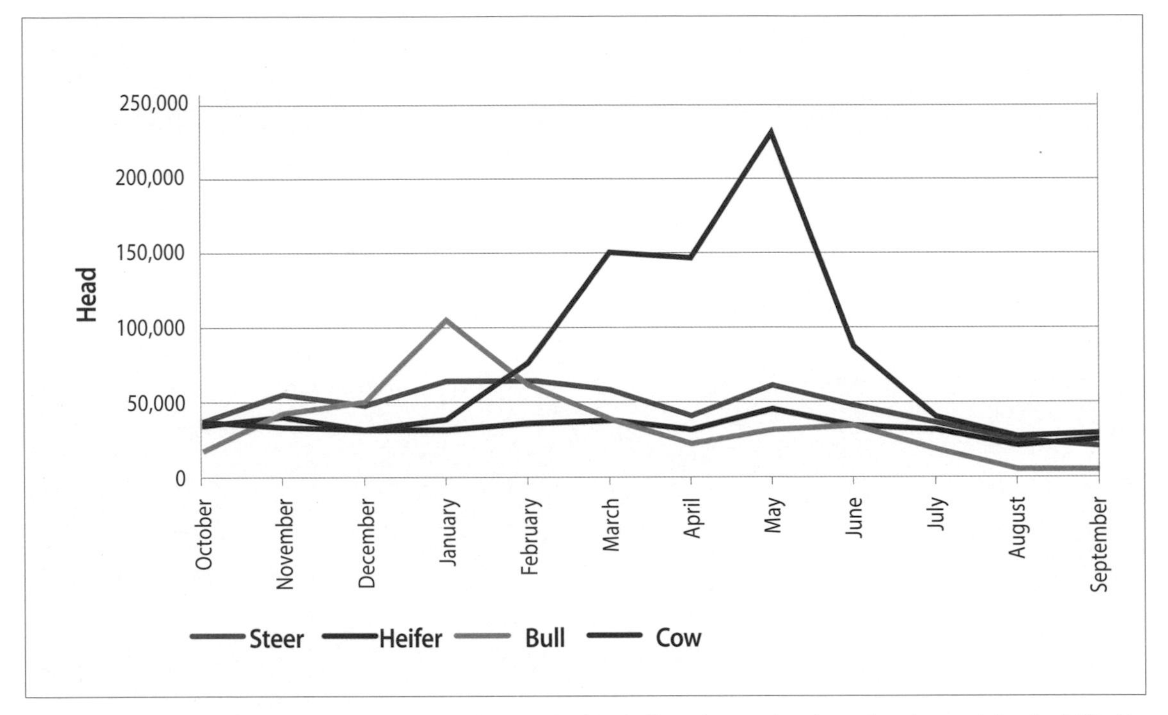

Figure 2.1 Annual slaughter patterns: numbers of steers, heifers, bulls and cows slaughtered each month for the 2013–14 season. Source: Beef & Lamb New Zealand Economic Service.

of age at a carcass weight of 220–240 kg, and are usually destined for the domestic market (although heifer beef is exported).

Beef cattle production is a seasonal industry, principally because most beef is produced from grass, and the animals sourced from the dairy industry follow a seasonal calving pattern. The major constraint to a consistent, year-round throughput of beef is the marked drop in availability of animals in the August to October period. This slaughter pattern in export processing plants is most pronounced for Holstein Friesian bulls but is also evident for steers and heifers. The marked seasonality in supply of stock (86–90 per cent of slaughter stock occurs during the months of November to June inclusive) is seen as a limitation on the ability of marketers to meet the demand of some markets. Their upper limits to supply become fixed by the availability of suitable beef in August, September and October.

Annual slaughter patterns for steers, heifers, bulls and cows for the 2013–14 seasons are shown in Figure 2.1. This slaughter pattern in export processing plants is most marked for cull dairy cows with increased levels in March and April building to a peak in May each year. Bull beef production reaches a summer peak in January to February from bulls grown to weight on spring grass. There is a second smaller autumn bull beef production peak in May to June before winter. The steer production pattern follows a similar but less pronounced trend to bull beef. Heifer slaughter is steady throughout the year, showing a small peak in May, and is linked to culls from the dairy herd.

Beef herd sizes are highly skewed because of the many small holdings, such as lifestyle blocks, that run a few beef cattle. Figure 2.2 shows that small holdings make up the majority of farms with beef cattle; however, these constitute a

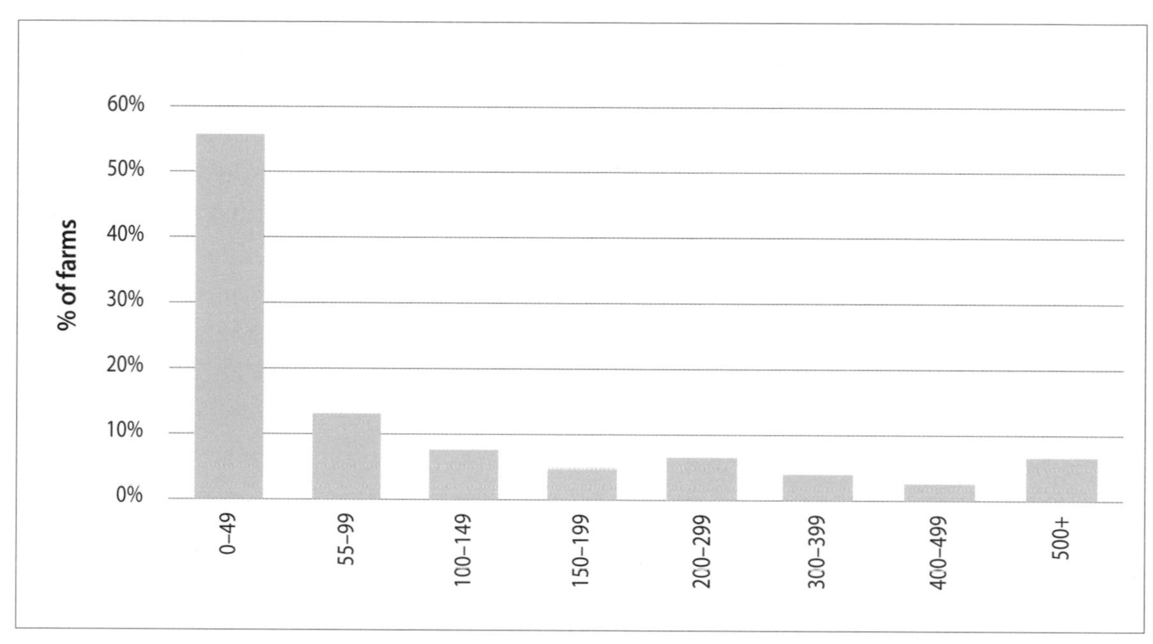

Figure 2.2 Beef cattle herd size distribution by size of holding, June 2013.
Source: Beef & Lamb New Zealand Economic Service/Statistics New Zealand.

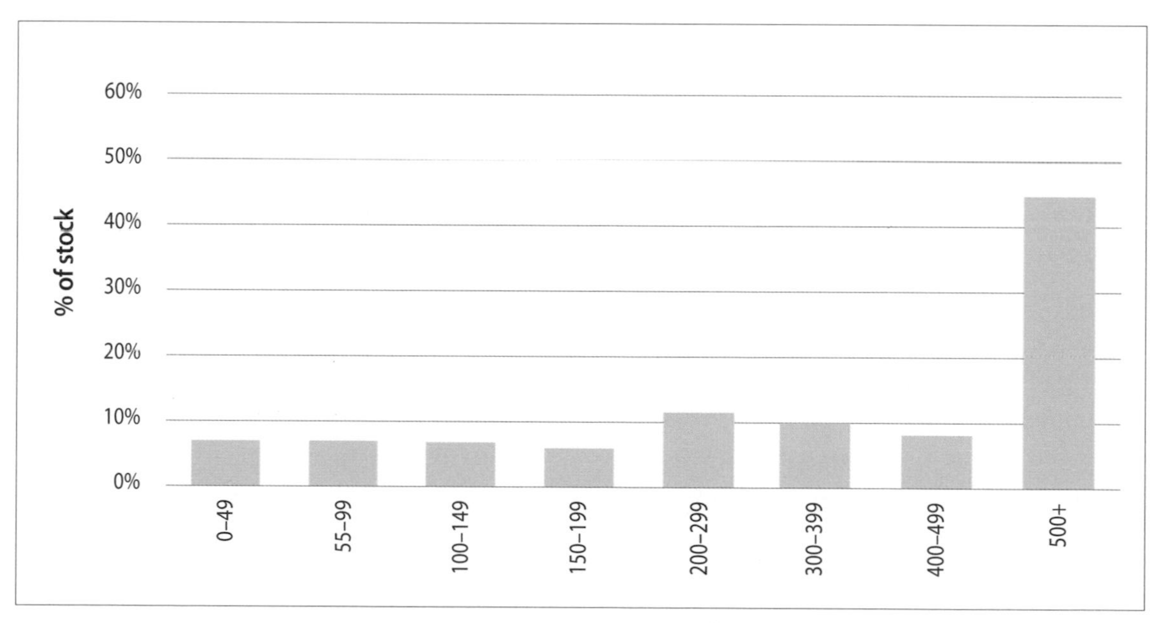

Figure 2.3 Beef cattle herd size distribution by proportion of total stock, June 2013.
Source: Beef & Lamb New Zealand Economic Service/Statistics New Zealand.

Table 2.7 **Beef cattle numbers by local region (at 30 June 2013)**

Region	Number of beef cattle ('000s)	% of total cattle
Northland/ Waikato/Bay of Plenty	1240	34
East Coast	936	25
Taranaki/ Manawatu	445	12
North Island	2621	71
South Island	1077	29
New Zealand	3,699	100

Source: Beef & Lamb New Zealand Economic Service.

relatively small proportion of the total beef herd. For example, 55 per cent of the beef holdings in New Zealand have fewer than 50 beef cattle each.

In aggregate, these holdings comprise just 7 per cent of total beef cattle. This group of farms is likely to be less responsive to industry conditions than the larger, more commercial farms. At the other extreme, 7 per cent of New Zealand farms have over 500 beef cattle each. In aggregate, these farms hold 45 per cent of total beef cattle, as shown in Figure 2.3.

About 71 per cent of New Zealand's beef herd is located in the North Island. While relatively evenly distributed throughout the North Island, the Northland/Waikato/Bay of Plenty region has 34 per cent of the total herd. Table 2.7 lists the major beef-cattle-producing regions.

A change in cattle numbers has recently been occurring in the lower part of the South Island, where substantial numbers of dairy-beef calves are now being sourced from the increasing number of dairy farms in the region.

Management of beef cattle

Beef cow calving coincides with the onset of the spring flush of pasture growth, thus ensuring adequate feeding levels post-calving and thereby encouraging maximum beef cow milk production. Calving on easy (or improved) hill country occurs in August, and on hard (unimproved) hill country in late September or October. Calf growth rates of 0.8–1.0 kg/head/day are usual while calves are suckling. This requires the provision of suitable pasture for rapid calf liveweight gain (typically a pasture mass of greater than 1500 kg of dry matter per hectare [kg DM/ha], or a 6–8 cm sward surface height) and to ensure that cows produce a calf each year (Morris, 2007).

Weaning usually occurs in March or April, when the calves are approximately 200 days old with liveweights ranging from 180–250 kg. Calves are usually castrated at three months of age and may be vaccinated and dehorned at this stage if necessary. They may be vaccinated again and de-wormed at weaning.

Cows are pregnancy tested in April after mob-mating where herds of 100 cows are run with three bulls. While this practice is common, single mating mobs, each with a ratio of one bull to 30–40 cows, is also a common practice.

The various classes of cattle, cows and calves, heifers and steers are kept separate and are moved around the farm most of the time with sheep. Weaned calves, if they are not sold in the autumn weaner sales, are given preferential treatment and are wintered on saved grass (paddocks where grass has been allowed to accumulate). Post-weaning feeding priority can be given to other classes of livestock, and after weaning cows become one of the few stock classes available on farm that can be fed restricted rations.

Table 2.8 **Target seasonal liveweights for beef breeding cows under various conditions**

	Liveweight (kg)			
	Weaning	Mid-winter	Pre-calving	Mating
Hill country (Angus)	450	400	420	430
Easy hill country (Angus x Hereford)	500	450	460	480
Easy/flat country	550	480	500	520

Source: Beef & Lamb New Zealand Economic Service.

This is a major justification for maintaining a breeding cow herd on hill country. Not only does this have significant advantages for the farm as a whole, but it has also been shown to be beneficial for the cows to lose around 10 per cent of their liveweight in the post-weaning period.

Some farmers rotationally graze their cows behind ewes in a winter rotation. In such situations, cow feed intakes are kept very low; for example Angus cows can eat as little as 3–3.5 kg DM/head/day. This highlights their efficiency and supports the contention that an efficiently managed beef cow could have a true winter stock unit cost of 3.5 stock units (1 stock unit is equivalent to 1 adult ewe).

The ability of adult beef cows to lose weight through the winter is dependent on body condition. Obviously, cows that are thin at weaning (a BCS of less than 5 on a scale of 1–10) — a situation that might arise after a prolonged summer drought — cannot be fed below maintenance level for extended periods. Table 2.8 gives typical seasonal target weights for breeding cows run under various environmental conditions.

The national calf weaning percentage (the number of calves weaned as a percentage of cows mated) is between 80 and 85 per cent, with little change over the past 20 years. Breeding heifers at 15 months of age is sometimes practised on properties where the heifers are grown to satisfactory liveweights at mating.

Heifers must achieve a critical breeding liveweight such that they are all cycling when the bull is put in with them. This will vary depending on the breed, but should result in an in-calf rate (percentage of cows pregnant) of 84 per cent in a mating period of 42–45 days. Good post-calving nutrition of heifers is even more critical than in adult cows, as heifers are more difficult to get back in calf. The choice of sire for first-calving heifers is also an important issue, with ease of calving being the major concern.

Angus bull.

Calendar of beef farming operations

This calendar is intended only to serve as a guide in describing the integration between beef and sheep farming systems. It is hypothetical because of the variety of districts, farming systems and stock performances involved.

JUL

This is often the beginning of the financial year. It is the time of the lowest soil temperatures and pasture production, and the wettest soil conditions of the year. Pugging damage of pasture by cattle should be watched for over the winter months. Breeding-cow feeding levels should be increased on saved pasture, hay, etc., in last six weeks before calving. The purchase of new breeding bulls should be considered.

AUG

Soil temperatures and pasture production are now increasing but soils are still wet. Breeding cows can be set stocked on saved calving padocks (pasture height 8–10 cm). Calving and lambing may compete for labour. Mature steers should be on saved pasture if they are to be sold in early spring, as otherwise liveweight gains will be low.

Pasture production may catch up with stock requirements, although this may not occur until mid or late October in some areas. Cows calve during the August to October period. Steers can be in the yearling sales (store cattle) at this time. There is a need to start improving heifer feeding levels as they approach mating. Bulls should be checked for fertility status.

OCT

Pasture production reaches its peak at the end of October and during November in many areas. Cows should be fed at high pasture levels. Bulls are joined (for a maximum of three cycles) with cows and heifers in early-pasture-growing districts. Steers and cull two-year-old heifers may be sold in the sale yards as store animals. Summer forage crops, if grown, are sown from October to January. Hay and silage paddocks are closed, dependent on pasture growth.

NOV

Spring pasture growth is now at its maximum. Bulls are joined with cows from October to January (maximum of three cycles). Calves are marked at three months of age and male calves are castrated. Spring cattle sales of steers, heifers and dry cows occur now. Hay paddocks can continue to be closed to control pasture growth.

DEC

Low-rainfall areas can begin to dry out, requiring water supplies to be checked from December through to March — especially when cattle numbers are high. Bulls may be rotated between mobs of cows (in case one is infertile).

Pasture production may be limited in low-rainfall areas. Bulls should be removed from breeding cow herds.

This is often a dry month with poor pasture production. Paddocks where hay has been cut and crop residues can be available for grazing. Sale of finished steers takes place. Water supplies must continue to be checked.

Pasture production may increase following autumn rains. Calves may be weaned. Pregnancy is diagnosed in cows and heifers. Surplus and cull cows, along with surplus weaner steer and heifer calves, are sold. Weaner calves are put on a drenching programme for internal parasites.

Autumn growth can be saved for winter or early spring feeding. Weaner calf sales occur this month. Cows can be restricted to a maintenance diet as they can lose weight over early winter, and can be used to clean up rank summer pasture growth. Weaner cattle should be grazing on saved feed. Pre-calving feed for cows must be planned, with paddocks closed up to allow pasture growth to accumulate.

Pasture production declines as soil temperatures cool. Stock numbers should be reduced to winter stocking rates. Feeding of winter forage crops to all classes of stock may begin. Cows are on maintenance feed rations, and can still be used to clean up rank pasture.

Pasture production is falling. The feeding levels of cows must start to increase towards the end of June by feeding on saved pasture. Bulls are purchased as rising two-year-olds.

Angus bulls.

Genetic improvement

Changes have occurred in the genetic make-up of the New Zealand beef cattle herds through the importation of new beef breeds to meet industry demand for improved growth and meat production. These imported European beef cattle breeds (e.g. Charolais, Simmental) have been crossed with the established breeds to obtain the desired traits and animals that are productive in the New Zealand environment. Some composite breeds (e.g. Stabilizer) have been developed.

Beef herds can be divided into a bull breeding segment (about 7 per cent of the population) and a non-breeding segment (93 per cent of the population), the latter usually being referred to as the commercial herds. Genetic flow from the bull breeding herds occurs mainly through natural mating (approximately 98 per cent), with the remainder by artificial insemination. Technologies based on DNA screening are sometimes used; for example, parentage testing is compulsory in some breed societies. Technologies using gene markers are in their infancy, although there are some genetic markers for meat tenderness, feed efficiency and marbling. The incorporation of these into traditional selection indices and estimated breeding values is also in its infancy and the industry awaits its development.

The Angus is the most important breed numerically. Although slower-growing than the European breeds, it matures early and gets in calf at a younger age with higher pregnancy rates. The productive and reproductive performance of Angus, Hereford and Angus x Hereford crosses under hill-country conditions makes these breeds more popular than other breeds of cattle. The New Zealand national beef herd comprises 23 per cent Angus, 11 per cent Hereford and 11 per cent Hereford x Angus. In addition, Angus and Hereford crosses with other breeds also contribute to a group classified as mixed crosses (36 per cent), while Holstein Friesian (12 per cent) and others (7 per cent) make up the rest. The heavier European breeds, especially Simmental, Charolais and Limousin, have made an impact as terminal sires, where all progeny (both male and female) are sold for processing or finishing. There has also been increased use of the Hereford x Holstein Friesian as a beef cow, with the Holstein Friesian introducing genes for higher milk production.

The expanding New Zealand dairy cattle industry represents a huge opportunity to produce surplus calves (both male and female) for the beef industry. Surplus capacity in the dairy industry could be increasingly used to produce more efficient beef suckler cows such as Hereford x Holstein Friesian or Angus x Jersey. The Angus x Jersey cross is an example of a smaller beef cow with high milk levels to produce a large calf at weaning and get back in calf quickly. These crossbred dairy-beef cows are often mated to Simmental, Charolais and Limousin bulls as terminal sires, as replacement cows do not have to be bred on farm.

Recent research has shown that there is little difference in meat eating quality between the breed types above. Contrary to popular opinion, beef with dairy cattle content is not inferior in eating quality to traditional beef breeds. However, there is some evidence to show that the fat from some cattle of dairy origin may be more yellow in colour.

Angus cows and well-grown calves on hill country.

Young Friesian bulls being grown for beef on a clover herb sward.

Nutrition

Sheep and beef cattle farms increasingly tend to be located in steeper hill country that is often of lower soil fertility and in many cases located in summer-dry regions. From a management viewpoint, sheep and beef cattle farms are relatively complex, with the same pastures having to meet several different feed requirements — including feeding ewes and beef cows, finishing lambs and growing cattle for slaughter.

Sheep generally graze pasture to a shorter residual height than cattle; hence, grazing policies are not consistent throughout the year but instead vary between seasons. For example, the same paddock may be set stocked or continuously grazed during spring, then rotationally or shuffle grazed at other times of the year.

Beef finishing systems

On intensive lowland farms, a common stock policy is to finish or fatten cattle. Store cattle are brought in as weaners in the autumn, or as one- to two-year-old store heifers or steers. Some cattle-finishing systems are run on farms where there is also a breeding cow herd.

Considerable flexibility and variation exist in the management objectives and feeding strategies for growing cattle. High rates of liveweight gain (i.e. over 1.0 kg/head/day) are required, so competition from sheep is avoided as far as possible. The cattle are often run on pasture (in excess of 1200 kg DM/ha or a 6 cm sward surface height) without other stock, and under this management practice liveweight gains can exceed 1.0 kg/head/day in the spring or around 0.5–0.75 kg/head/day during the summer and autumn seasons.

A form of specialist finishing has arisen in New Zealand over the past 30 years, based on Holstein Friesian bull calves sourced from the dairy industry. These bulls are normally slaughtered between 18 months and two years of age. The calves are either sold at four days of age direct to specialist calf-rearers or to sheep/cattle farmers, or are reared on dairy farms to three months of age and then sold in the October to November dairy-beef weaner sales. Beef from Holstein Friesian bulls is used primarily as lean processing or ingredient beef (Cosgrove et al., 2003; McRae, 1988, 2003).

There are many different classes, ages and conditions of cattle that can be purchased for finishing. The particular market that a farmer aims for will determine the type of cattle purchased. For example, a farmer wishing to supply the local (domestic) trade might choose early-maturing cattle, such as Angus heifers, with typical carcass weights of 220–240 kg. In comparison, a farmer wanting to supply the North American processing beef trade would probably choose Holstein Friesian bulls and finish to carcass weights of 270–320 kg.

Another major determinant of the type of beef finishing system used is the schedule price at the proposed time of sale: a farmer may choose to base the finishing system around the supply of animals in a particular month when, historically, schedule prices are high (e.g. July or September).

Beef carcasses in the chiller.

Disease

Bovine tuberculosis

Bovine tuberculosis (bTB) is potentially one of New Zealand's most serious animal health problems. Tuberculosis in cattle is an infectious disease caused by the bacterium *Mycobacterium bovis* and can be transmitted to humans. Bovine TB can be a major health problem in deer as well as cattle; it also infects a wide range of wild animals, especially possums, wild deer, ferrets and pigs. Tuberculosis is spread mostly by close contact between animals; infection is usually caused when one animal inhales bacteria coughed up or exhaled by another. In New Zealand, livestock contract bTB mostly from contact with infected wildlife, although infection can occur within herds between animals. Tuberculosis can be controlled by controlling the disease both in livestock and in the wild vector population. Less than 0.2 per cent of New Zealand's cattle and deer herds are now infected, following progress with the control programme. The objective now is to move from control to eradication of the disease.

Disease control in cattle herds involves:

- bTB testing of at-risk herds and slaughtering of animals that react positively to the test
- classifying herds by their bTB status
- controlling the movement of animals from infected herds or areas where herds are at greater risk of a bTB breakdown

Vector control involves reducing the populations of vectors (usually possums, but sometimes ferrets and wild deer) by a combination of trapping, poisoning or shooting.

All cattle herds are classified according to their history of bTB infection. The classification is designed to provide a tool for measuring the risk of bTB infection in any herd of animals. Essentially, there are two classifications:

- A herd will be classified as infected (I) if bTB has been confirmed by testing, post-mortem or another approved diagnosis. A newly infected herd will be classified as I1, with a number (up to 10) added for every year it remains infected — thus, a herd that has remained infected for three years will be classified as I3.
- Herds which show no evidence of bTB from tests or other diagnoses will be classified as clear (C). Again, a number (up to 10) will be added for each year that the herd has been confirmed by testing to be clear of infection.

For a herd to shift from infected to clear 1 (C1), it must have two consecutive clear whole-herd tests with a minimum of six months between the tests and with no further evidence of disease. Under normal circumstances it will take at least one year for a herd to change from infected to clear status.

Animal status declarations

All movements of cattle and deer one month of age or over must be accompanied by a completed animal status declaration card. This includes movement to slaughter. All of the questions on the declaration card must be answered fully and correctly.

Angus bull.

National Animal Identification and Tracing (NAIT)

To assist with traceability of cattle in the marketplace, all cattle must be tagged with NAIT-approved RFID tags (radiofrequency identification tags). These animals must be tagged within six months of birth or before they move off their farm of birth — whichever is soonest. NAIT tags are the only tags required by law.

The only exemptions to the tagging requirement are for calves less than 30 days of age going direct to a meat processor (bobby calves) with a direct-to-slaughter tag issued by the meat processor, and for cattle considered too dangerous to tag that are also going directly to a meat processor.

Every farm with beef cattle needs a NAIT identification number even if a tagging exemption applies to its animals. When a farmer moves beef cattle off farm or receives them on farm, the sending movement must be recorded or the receipt of animals confirmed in the NAIT system. This includes movements for private sales, grazing stock and service bulls. If a farmer sends cattle to a NAIT-accredited sale yard or meat processor, these will record the movement for the farmer. Farmers receiving animals from a NAIT-accredited sale yard must confirm the receipt. All movements of cattle and deer one month of age or over must also be accompanied by a completed animal status declaration card.

Internal parasites

Gastrointestinal parasites (worms) are the most important production-limiting disease of young beef cattle. The main worm species of significance in New Zealand cattle are *Ostertagia ostertargi*, *Trichostrongylus axei* and *Cooperia oncophora*. The current parasite control programmes rely on the regular use of anthelmintics (de-worming medication). Grazing pressure, stocking rate and pasture type can all influence the rate at which young cattle can become infected with larvae from pasture.

It is in cattle in their first year of life that the effects of parasitism are likely to be the most severe and of the greatest long-term significance. That is not to say that such effects are confined to this age group — but these animals have little or no acquired resistance, have minimal reserves on which to draw, and have a greater need to be making maximal skeletal and muscular growth if their lifetime production is not to be impaired.

The principal objective of all control strategies should be to minimise the exposure of cattle to infectious larvae on pasture. It is key to control the levels of parasitism in the first 12–15 months of life, not only because animals of this age are the most vulnerable but also because they are potentially the main source of pasture contamination, which has a direct bearing on the risk of exposure for older animals.

Obviously, the control measures taken need to be parasitologically effective as well as cost-effective. However, consideration must also be given to their compatibility with market requirements with respect to such issues as residues, and to sustainability in the longer term.

Parasite control measures in beef cattle are heavily dependent on the use of anthelmintics. Many farmers growing young cattle use a four- to six-week preventative or insurance type drenching programme. Others use a visual or best-guess method, or drench when they yard animals to weigh them. All of these approaches often miss a parasite problem until a production loss occurs, and are becoming unsustainable due to a number of factors. Because of their ease

Well-grown Angus calves and cows.

of use ('pour on' or injection), it has been estimated that 80 per cent of the anthelmintics used on cattle come from a single category — macrocyclic lactones (MLs), such as ivermectin (sold under the brand name Ivomec), doramectin (Dectomax), moxidectin (Cydectin, Vetdectin), eprinomectin (Eprinex) and abamectin (Genesis). This heavy use of one category of drench has created resistance problems; notably, these MLs are not particularly effective in controlling *Cooperia*.

From first principles, a 'monoculture' of young cattle on a continuous grazing system is a recipe for disaster from a parasite control perspective. Experience with beef units that have been running for a number of years has confirmed that this is the case. The current recommendation would have to be that, realistically, a system running only rising yearling bulls is not sustainable on a long-term basis from a parasite management perspective. A split policy that included rising two-year-old bulls could be used, but there are many other classes of stock that can be integrated into these units which are more beneficial from a parasite control perspective (e.g. ewes/cows).

Environmental considerations

The major environmental issues faced by the sheep and beef cattle industries revolve around water quality and supply, climate change and greenhouse gas (GHG) emissions. Sheep and beef cattle contribute significantly to the economic wellbeing of New Zealand, therefore minimising environmental impacts while improving animal performance and economic returns is a major national concern.

Most of the sheep and beef cattle in New Zealand are farmed on hill or high country at low stocking rates and under less intensity than the dairy cattle industry. However, farming to reduce the environmental footprint is still seen as an emerging feature of the operating environment for sheep and beef cattle in New Zealand.

Leaching of nitrogen and loss of phosphorus are undesirable effects of agricultural intensification, as they are considered to be pollutants in waterways. Ruminants typically excrete 75–95 kg of the nitrogen in their diet over a year, with excess dietary nitrogen excreted mainly in the urine. Most nitrogen is leached during winter and spring, when rainfall exceeds evapotranspiration and soil moisture status is high.

Hereford cows cleaning up rough pasture.

Current research into mitigation measures to reduce nitrogen leaching has focused on the amount of time the animals spend on pasture or the duration of controlled grazing on winter crops. Most of these strategies are used or being trialled on dairy farms, with few being trialled on hill-country sheep and beef farms. New Zealand's extensive low-cost sheep and beef farming systems mean that housing to control the potential environmental effects is simply not an option. There are, however, some advances to be gained from the precision application of fertiliser by aeroplanes, and this is an active area of research in New Zealand.

Ways of reducing the amount of potential pollutants entering waterways on sheep and beef cattle farms, especially in hill-country properties, pose significant challenges for the industry. The exclusion of cattle by fencing them off from the many creeks and waterways on hill-country farms may be too costly for some farms. Nevertheless, in the future all farm types are likely to be evaluated for nitrogen losses, and the nitrogen loss restrictions currently being applied to dairy farms will increasingly be applied also to sheep and beef farms. The development of all future farm systems is likely also to need an environmental impact evaluation if such systems are to be accepted by society.

Friesian bulls for bull beef production on hill country.

Animal welfare

There are few major animal welfare issues associated with traditional beef cattle production in New Zealand. These animals graze mostly on extensive properties with opportunities to move and seek shade and shelter as needed. They are well capable of dealing with winter weather and can eat rough, poor-quality feed. Pasture shortages during droughts or winter are usually countered with hay or baleage. Beef cattle are yarded infrequently for weaning, vaccination, pregnancy testing and a few other procedures.

Bull calves from beef cows are usually castrated and are dehorned if necessary. Many beef calves are polled (born without horns). Dairy bull calves reared for bull beef are not castrated and are usually disbudded early in life. Calves born to beef cows are weaned at about six months of age; while this is stressful, good management can reduce its significance. Dairy bulls reared for beef have to be managed to minimise fighting and injuries. Groups are formed before puberty, and to minimise social disturbance no new animals should be added to existing groups.

Beef cattle are slaughtered under strict veterinary control. Cattle are examined before slaughter to ensure that they are fit to be transported and for slaughter. All animals are stunned before being bled out (exsanguinated).

Of all the animals farmed in New Zealand, beef cattle have the fewest serious welfare issues. There are few feedlots, cattle are generally transported over short distances, and no cattle are shipped overseas for slaughter. The welfare of steers or heifers born on a sheep and beef farm and finished either there or on a finishing farm is probably as close to ideal from a welfare perspective as the farming of animals allows.

Conclusion

New Zealand beef cattle farmers can look to the future with some confidence, as they produce a high-quality grass-finished product that uses relatively little energy compared with feed-lot-reared cattle in the northern hemisphere. Issues that farmers need to face in the future might include the reliance on the quota allocation for beef to the USA of around 50 per cent of New Zealand's beef exports, and therefore the need to diversify into other high-priced markets such as North Asia. Diversification will require considerable skill in negotiation between producers, processors and exporters to overcome seasonality of supply and to guarantee traceability and a high-quality product delivered to specification and at the right time.

References

Beef & Lamb New Zealand. (Various years). Annual Reports. Wellington: Beef & Lamb New Zealand.

Campbell, A.G. (Ed.). (1970). *New Zealand beef: production, processing and marketing*. Wellington: New Zealand Institute of Agricultural Science.

Cosgrove, G.P., Clark, D.A., & Lambert, M.G. (2003). High production dairy-beef cattle grazing systems: a review of research in the Manawatu. *Proceedings of the New Zealand Grassland Association*, 65, 21–28.

McRae, A.F. (1988). Bull beef production. The Tuapaka experience. *Proceedings of the New Zealand Grassland Association*, 49, 41–45.

McRae, A.F. (2003). Historical and practical aspects of profitability in commercial beef production systems. *Proceedings of the New Zealand Grassland Association*, 65, 29–34.

Morris, S.T. (2007). Pastures and supplements in beef production systems. In P.V. Rattray, I.M. Brookes & A.M. Nicol (Eds), *Pasture and supplements for grazing animals* (pp. 243–54), Occasional Publication no. 14. Hamilton: New Zealand Society of Animal Production.

Morris, S.T., & Kenyon, P.R. (2014). Intensive sheep and beef production from pasture — a New Zealand perspective of concerns, opportunities and challenges. *Meat Science*, 98, 330–35.

Angus cows and calves.

Chapter 3

Sheep Production

Lydia Cranston, Anne Ridler, Andy Greer and Paul Kenyon

Chapter 3

Sheep Production

Lydia Cranston, Anne Ridler and Paul Kenyon
Institute of Veterinary, Animal and Biomedical Sciences
Massey University, Palmerston North
Andy Greer
Lincoln University, Canterbury

Introduction

The sheep industry has been a cornerstone of the New Zealand economy since the nineteenth century (Carter & Cox, 1982), and it has been stated that New Zealand's economy developed 'on the sheep's back'. During the Korean War in the 1950s, the price paid to New Zealand farmers for coarse wool, adjusted to 2015 prices, was $64/kg (Anon, 2013). Today's coarse wool prices are nowhere near that value, but sheep, which are usually farmed in conjunction with beef cattle, still dominate land use. They are farmed on nearly two-thirds of New Zealand's agricultural and forestry land, on more than 40 per cent of farms (Ministry for Primary Industries, 2012).

Sheep numbers have declined from a peak of almost 70 million in the early 1980s to just below 30 million in 2015 (see Figure 3.1). However, despite this decrease, the total production of sheep meat has been maintained, due to significant increases in sheep production with little change in animal stocking rates (Morris & Kenyon, 2014). There was a 23 per cent increase in lambing percentage from less than one lamb per ewe (98 per cent) in 1987 to 121 per cent in 2013. In addition, average carcass weights of lambs have increased from 14 kg to 18 kg (Morris & Kenyon, 2014). Bray (2004) has estimated that this rise in carcass weight was driven by a 50 g/day improvement in lamb growth rate.

Today most New Zealand sheep are Romneys,

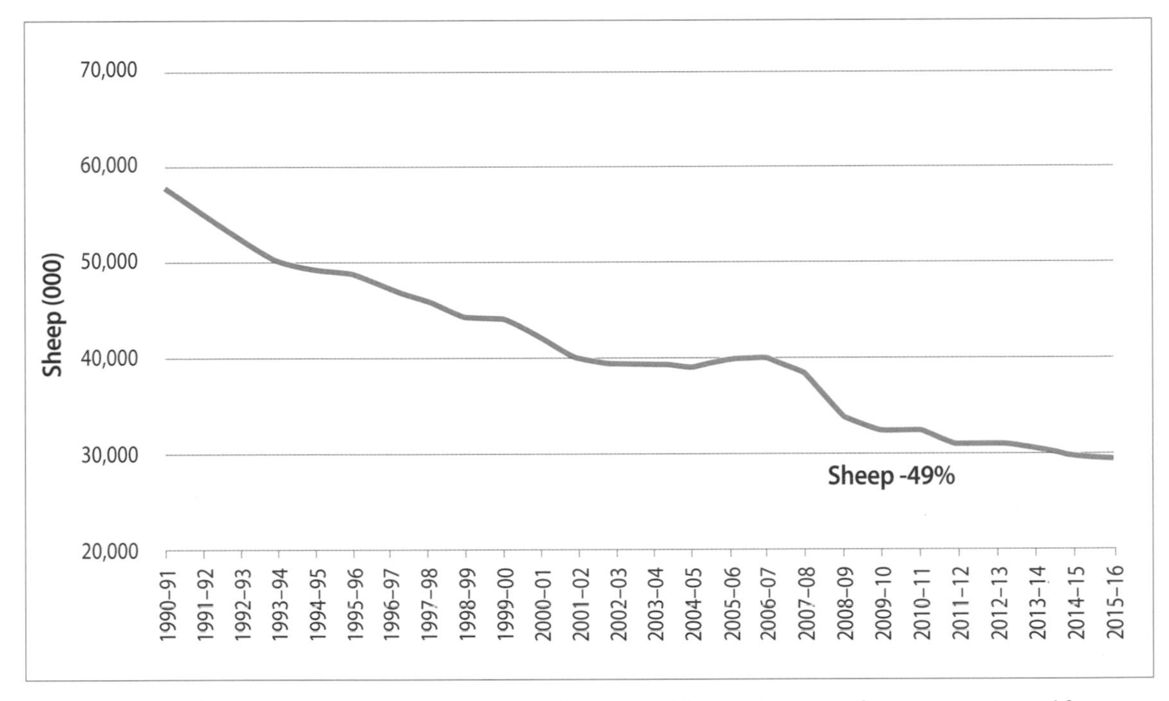

Figure 3.1 Change in the number of sheep in New Zealand between 1990–91 and 2015–16 (figures are estimated for 2015–16). Source: Beef & Lamb New Zealand Economic Service.

their breed derivatives (Coopworths, Perendales) and a wide range of 'modern' composites based on these. These are dual-purpose breeds producing lambs for meat as well as coarse wool, the latter often used in carpet manufacture. Since the early 1990s there has been increased use of two-breed, three-breed and four-breed composite sheep, generally with Finnish Landrace, Texel and East Friesian breeds' genes being incorporated into an existing Romney, Coopworth or Perendale flock. A recent survey has suggested that more than 40 per cent of sheep flocks include composites (Corner-Thomas et al., 2013). Fewer than 10 per cent of sheep are Merino or Merino-based breeds, which produce fine wools suitable for the manufacture of clothing.

A new development in the sheep industry is sheep milking. This will require a degree of industry coordination, investment and research both pre-farm-gate and post-farm-gate to ensure it does not become a boom-and-bust industry.

Romney-type lambs. **Previous:** Romney-type ewes.

The New Zealand sheep industry

The sheep industry has a strong export focus, with 95 per cent of sheep meat and 90 per cent of wool being exported (Morris, 2013). In the year ending 30 September 2014, New Zealand produced an estimated 486,000 tonnes of sheep meat and 167,000 tonnes of wool (Beef & Lamb New Zealand, 2016a). The country is the largest exporter of lamb meat in the world, accounting for nearly half of the global trade in lamb. The largest market for lamb from New Zealand, by volume, is the EU, with the major importing countries being the United Kingdom, France and Germany (Table 3.1); the single largest importing country, by volume, is China. Other major markets include the North Asia region (Japan, South Korea and Taiwan) and North America. On a price per kilogram basis, North America and the EU are the most valuable.

Sheep meat is classified as either lamb or mutton. Lamb is meat from sheep that are less than about one year of age; this age group is identified by having deciduous teeth and no permanent incisors. When the first two permanent incisors erupt, lambs are classed as two-tooths and their carcass value declines in value immediately. Mutton is sourced from all sheep not categorised as lambs. The export markets for lamb and mutton vary, with significant volumes of mutton going to North Asia, in particular China (Table 3.1).

Table 3.1 **Summary of the major sheep meat (lamb and mutton) export markets for New Zealand, for the year ended 30 September 2015**

Export market	Lamb (% of total tonnes)	Mutton (% of total tonnes)
Africa	1	1
European Union	41	15
Middle East	10	2
North America	10	8
North Asia	32	64
Pacific	2	0
South Asia	2	8
Other	2	2
Total	**100**	**100**

Source: Beef & Lamb New Zealand Economic Service.

The New Zealand slaughter industry comprises four major processing and export companies (AFFCO, Alliance, ANZCO and Silver Fern Farms), which together account for approximately 75 per cent of New Zealand's sheep meat processing. These processors each operate multiple slaughter plants, located throughout the country. There are also a number of smaller processing plants, which generally service niche markets.

In the 1970s, New Zealand sheep meat was predominantly exported as whole frozen carcasses with a small proportion of frozen cuts. Since then there has been a shift away from exporting whole frozen carcasses towards exporting value-added products, including chilled meat, frozen cuts and boneless frozen cuts. These consumer-ready cuts and chilled products achieve premium prices over frozen carcasses.

On a clean wool basis, New Zealand is the fourth-largest producer of wool in the world, producing just over 10 per cent of total global production (Beef & Lamb New Zealand, 2016a). More than 75 per cent of this wool is described as crossbred wool type, with a fibre diameter

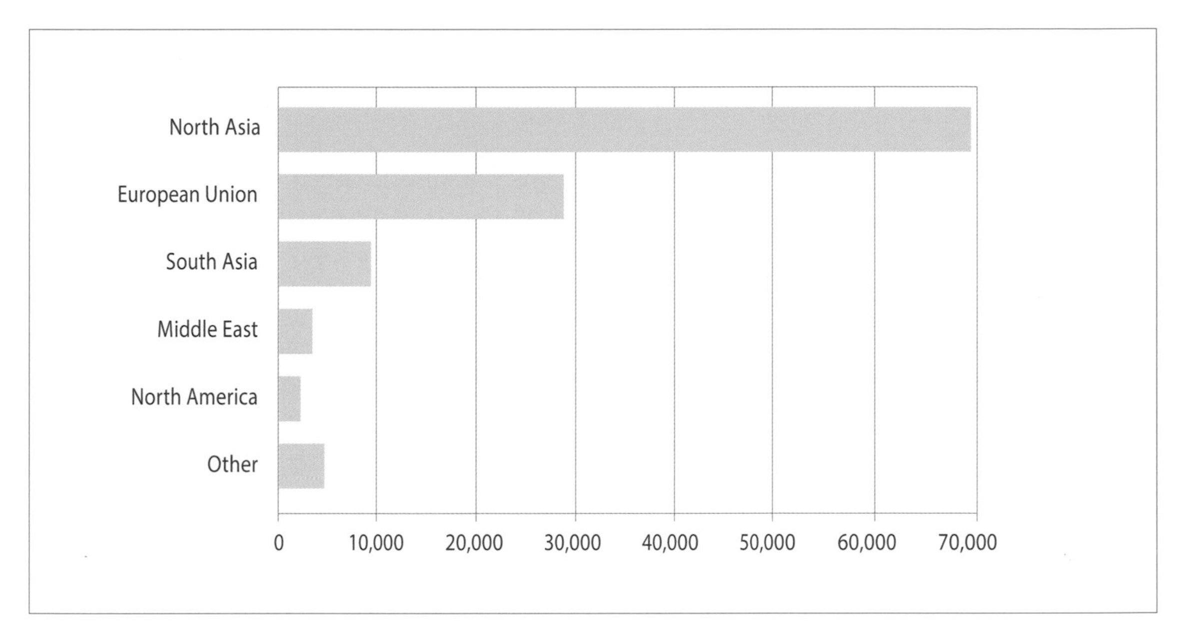

Figure 3.2 Summary of the major wool export markets (tonnes of clean wool) for New Zealand, for the year ended 30 June 2015. Source: Beef & Lamb New Zealand Economic Service.

of greater than 31 µm. It is mostly used in the carpet and coarse outer garment markets. Fine wool, with a fibre diameter of less than 25 µm, makes up approximately 8 per cent of New Zealand's production. Within breed type, lamb's wool is generally 2–3 µm finer than mature ewe wool and is commonly used for producing socks and finer outer garments. More than 10 per cent of wool is harvested as slipe wool (wool from the skins of slaughtered sheep) from the pelts (skins) of slaughtered sheep (often lambs). The vast majority of the wool clip is exported to North Asia and the EU (Figure 3.2).

In New Zealand approximately 40 per cent of all wool is sold off farm through private sales, while the majority is sold via auctions (Beef & Lamb New Zealand, 2016a). A relatively small amount of wool is sold by a farmer directly to a processor, and a smaller amount still is sold via a partnership between a grower and a processor where the final product is sold under the grower's brand.

New Zealand also has a newly developing, and small, sheep dairy industry — a response to the growing international demand for sheep milk and cheese. The building of infrastructure and the development of business networks is likely to result in further increases in the number of sheep milking operations in New Zealand. Current issues affecting the growth of the industry include a lack of processing capability, current New Zealand sheep breeds not displaying prolonged lactation periods, and the management and welfare of newborn lambs.

Glossary

Aged: sheep of any age over 5 years.

Cast-for-age (CFA): ewes culled from a flock on account of old age. These ewes are usually mated to prime-lamb sires for one or two years.

Crutching: removing the wool (and dags) from the groin and perineal regions.

Cryptorchid: a castrated male whose testicles have been pushed against the body wall or into the abdomen and the scrotum removed. Also known as short-scrotum.

Dags: clumps of faeces that adhere to the wool around the perineal region. Dagging refers to the removal of dags.

Dry ewe: a ewe that fails to get pregnant.

Fibre diameter: the average fibre diameter of a sample of wool. It is measured in microns, µm (1 µm = 1 millionth of a metre).

Flock: a number of sheep, usually a breeding group or unit on the farm.

Four-tooth: sheep aged 2–2¾ years, with four permanent central incisor teeth.

Four-year or full-mouth: sheep with eight incisor teeth; older than 3 years.

Hogget: young weaned sheep (aged 4–16 months).

Interbreeding: the mating together of two similar crossbreds such as (Cheviot x Romney) ewe x (Cheviot x Romney) ram.

Lamb: young sheep from birth in spring to about mid-autumn. There is no strictly defined age at which a lamb becomes a hogget.

Mixed age: mixed ages in a flock.

One-year ewe: an older ewe suitable for one year as a prime-lamb dam.

Prime lamb: this term replaces 'fat lamb' as market requirements have reduced the fat levels of lamb carcasses. The present objective of the prime-lamb breeder is a low proportion of fat to muscle.

Quality number or wool count: subjective appraisal of wool for fineness. The quality numbers normally used in New Zealand range from 36s (strong) to 70s (fine). This system has now been replaced by diameter.

Six-tooth: sheep aged 2½–3½ years, with six permanent central incisor teeth.

Staple: lock or tuft of wool fibres.

Store lamb: lamb that is sold to another property for finishing (growing to slaughter).

Teaser: a ram rendered non-fertile by vasectomy.

Tupping: another name for mating.

Two-tooth: sheep aged 15–24 months, with two permanent central incisor teeth.

Wet ewe: a ewe that has lambed and reared the lamb to docking or weaning.

Wet-dry ewe: a ewe that has carried a lamb to late pregnancy or birth (such that her udder has increased in size) but whose lamb has died before docking or weaning.

Wether: a castrated male whose testicles have been removed.

Romney-type ewes.

A brief history of sheep production

Sheep have played an important part in New Zealand's history over the past 150 years. Captain Cook's journals indicate that he landed sheep in Queen Charlotte Sound in 1774, and the missionary Samuel Marsden brought sheep into the Bay of Islands in 1814. However, permanent populations of sheep do not seem to have survived from these two importations. In 1834, Allan Bell Wright landed Merinos on Mana Island, off the Kapiti Coast, and wool exportation began the following year.

Wool became New Zealand's first farm-based export product, as it could be easily shipped over long distances with little deterioration. Early importations of Merino sheep were followed by introduction of British breeds, including the English Leicester, Romney Marsh, Border Leicester and Lincoln. Merinos continued to be the main breed imported, however, as Australia was a source of the breed and the type of tussock-land farmed at that stage suited the animals. In 1855 there were fewer than 1 million head but by the 1880s this had grown by more than tenfold. This enormous increase was made possible through large-scale importation of Merinos that were readily available from Australia.

In the 1860s, gold was discovered in the South Island. This led to a massive increase in the human population there and a corresponding increase in the demand for meat. The expansion of this market led to a further increase in sheep numbers and encouraged the development of meat rather than wool characteristics in the sheep of the time. During this period, considerable amounts of bush were cleared in both the North Island and the South Island with large areas converted to grass farming, increasing the demand for breeding ewes. However, much of this converted land was not suited to Merinos; wetness caused footrot and fleece problems. Rams from a variety of British breeds, including Romney Marsh, Border Leicester, English Leicester, Lincoln and Cotswold, were therefore crossed with Merino ewes to build up a national flock suited to wetter conditions. In the South Island, two types of crossbred became stable: the New Zealand Halfbred and the Corriedale. However, the Merino was still the main breed given that wool, not meat, remained the main export commodity.

A major development occurred in 1882 when the *Dunedin* sailed from Port Chalmers carrying frozen carcasses bound for England, heralding the start of the refrigerated meat export industry. Due to the poor meat characteristics of the Merino, the Romney then started to become popular, especially in the North Island, and by 1921 more than 70 per cent of the rams in the North Island were Romneys. The Border Leicester and the Shropshire were the main terminal-sire breeds, but in the 1920s the Southdown became a popular terminal sire because its crossbred progeny were small, compact and early finishing, which suited the UK market. This breed dominated the terminal market for the next 40 years. During this period Romney numbers increased and Merino numbers declined. The Romney was selected for a small, compact-shaped carcass and increased wool growth. Romney wool is coarse and best suited for carpets and coarse outer garments.

It became apparent in the 1940s and 1950s

Romney-type lamb with tail.

that the Romney was not performing to expectations. To improve flock performance on easier country, it was crossed with the Border Leicester to produce the Coopworth. This breed has greater milk production and fecundity than the Romney. On steeper country, crossing of the Romney with the Cheviot produced the Perendale, which is a slightly smaller animal than the Romney but is hardier and thought to be more suited to hill-country conditions. During this period the majority of farmer incomes still came from wool. Another distinct New Zealand breed, the Drysdale, was also developed at this time following the identification of a recessive gene in the Romney that resulted in a coarse wool fibre that was particularly suited for carpet wool production.

In the 1970s and 1980s, consumers started to become concerned with the level of fat in their diets. The fatter carcass of the Southdown became less popular, resulting in increased use of larger but leaner terminal breeds such as the Suffolk and the Poll Dorset. By the early 1980s sheep numbers had peaked at more than 70 million, dominated by the Romney and followed by the Perendale and the Coopworth. Combined, the Merino and the two other fine-to-medium wool breeds, the New Zealand Halfbred and the Corriedale, now made up less than 10 per cent of the national flock. During the 1980s, however, it become apparent that the value of Romney-type wool was on the decline, and that if farmers were to be financially viable they needed to produce more lambs per ewe with heavier carcasses. This encouraged the importation of a number of 'exotics' such as the Finnish Landrace, the East

Friesian and the Texel, which were chosen for their high fecundity, milk yield and/or carcass quality. Rather than using these as straight-bred animals, many farmers used them to produce composite breeds that displayed higher reproductive rates and faster lamb growth rates than the traditional Romneys, Coopworths and Perendales.

At this time, traditional ram breeders began to accept that if they wanted to stay viable they had to move away from selecting rams on the basis of wool traits and focus on the more economically valuable reproduction, growth and carcass characteristics. These breeders made significant changes to their breeds, and modern examples are very different from those existing 30 years ago.

During the late 1990s and the 2000s, much of the easier country traditionally used for sheep farming was converted to dairy cattle farming, orchards and viticulture. This has resulted in the national flock falling to around 30 million head, most farmed on hill country. There are now more than 100 breeds or composite types of sheep in New Zealand, which can be categorised as dual-purpose, terminal, fine-wool and milk breeds or types. The major breeds, along with their typical traits, are listed in Table 3.2.

Farm types

Sheep farms are classified into eight types (classes) based on a combination of factors such as topography, environment and farming system (Beef & Lamb New Zealand, 2016b).

- **Class 1:** South Island high country. Extensive run country at high altitude, carrying fine-wool sheep, with wool as the main source of revenue. Located mainly in Marlborough, Canterbury and Otago.
- **Class 2:** South Island hill country. Mainly mid-micron wool sheep, mostly carrying between 2 and 7 stock units per hectare. Generally, three-quarters of the stock units wintered over are sheep and one-quarter are beef cattle.
- **Class 3:** North Island hard hill country. Steep hill country and low-fertility soils, with most farms carrying 6 to 10 stock units (ewes) per hectare. While some lambs are finished for slaughter, a significant proportion are sold as store lambs.
- **Class 4:** North Island hill country. Easier hill country and/or higher-fertility soils than Class 3 soils; mostly carrying between 7 and 13 stock units (ewes) per hectare. A high proportion of lambs are sold for forward store or in prime condition for slaughter.
- **Class 5:** North Island intensive finishing farms. Easy-contour farmland with the potential for high production; mostly carrying between 8 and 15 stock units per hectare. A high proportion of lambs are sent to slaughter and replacement ewes and lambs are often bought in.
- **Class 6:** South Island finishing-breeding farms. A more extensive type of finishing farm, also encompassing some irrigation units and frequently with some crops grown. Carrying capacity ranges from 6 to 11 stock units per hectare on dryland farms and over 12 stock units per hectare on irrigated units. Mainly located in Canterbury and Otago. This is the dominant farm class in the South Island.
- **Class 7:** South Island intensive finishing farms. High-producing grassland farms carrying about 10 to 14 stock units per

Table 3.2 **Summary of traits of selected sheep breeds in New Zealand (note that there is significant within-breed variation and the values are guidelines only)**

	Descriptor	Ewe body weight (kg)	Lambing percentage	Fibre diameter (microns)	Fleece weight (kg)
Maternal breeds					
Romney	Medium–large. Most predominant breed.	55–70	110–150	33–40	4.5–6.0
Coopworth	Medium–large, moderately hardy. Easy lambing and good mothering and milking ability.	60–75	110–170	35–39	4.5–6.0
Cheviot	Compact, short-legged, erect pointed ears.	45–55	90–120	28–33	2.5–3.5
Perendale	Medium-sized, long body. Active, hardy.	50–65	100–140	31–35	3.5–5.0
Merino	Medium-sized, fine-boned. A speciality fine-wool breed.	40–55	75–110	16–24	3.0–6.0
Drysdale	Medium–large. Coarse long wool.	55–65	90–120	40+	5.0–7.0
Corriedale	Medium-sized. Suited to drier environments.	60–75	90–130	28–33	4.5–6.5
Terminal breeds					
Poll Dorset	Medium–large, pink nose and skin. Good milking ability. Can breed out of season.	60–80	120–170	27–32	2.0–3.0
Suffolk	Medium–large, black face and legs.	60–80	120–150	30–35	2.5–3.0
Southdown	Medium-sized, mouse-coloured face, short legs. Early maturing.	50–70	110–140	23–28	2.0–2.5
Texel	Medium-sized, black nostrils. Face and legs clear of wool.	50–65	110–140	33–37	2.5–4.0
Speciality breeds					
Finn	Medium-sized. Highly fecund.	50–70	175–250	27–30	2.5–4.0
East Friesian	Large, thin-tailed. High fecundity and good milk production.	75–95	160–330	35–37	4.0–5.0
Dorper	Medium–large. Self-shedding, fleece not used for commercial purposes. Can breed out of season.	65–80	110–130		
Wiltshire	Medium-sized. Self-shedding.	60–75	190–210	30–32	

Sources: These descriptive traits were developed with the support of data from *Pocket guide to the sheep breeds of New Zealand* (Southby, 2008), *Sheep breeds of New Zealand* (Meadows, 1997) and *New Zealand sheep and their wool* (Morrison, 1980).

hectare, with some cash crops. Located mainly in Southland, and South and West Otago.

- **Class 8:** South Island mixed cropping and finishing farms. Located mainly on the Canterbury Plains. A high proportion of farm revenue is derived from grain and small seed production as well as finishing lambs or beef cattle.

Sheep performance differs widely across these farm classes (Beef & Lamb New Zealand, 2016c). Each year, detailed data is collected for a subset of farms within each farm type; these are summarised in Table 3.3.

As a result of the recent increase in the dairy cattle industry in New Zealand, a greater proportion of sheep farms are now located on hill country, which renders the gains outlined earlier even more impressive. As indicated in Table 3.3, production varies widely across the eight classes. The relative change in the value of sheep meat and wool has led to wool production now making only a minor contribution to total income on most sheep farms, it being primarily driven by meat production.

The New Zealand sheep industry also comprises ram breeding farms (stud farms), which specialise in breeding for genetic improvement in the traits appropriate for their breed type. Commercial farms buy their rams from these stud farmers, achieving genetic gain in their flocks as a result of purchasing rams from breeders who share the same production aims.

Table 3.3 **Selected provisional commercial sheep and beef farm class production data for 2014–15**

	Beef & Lamb NZ Farm Classes							
	1	**2**	**3**	**4**	**5**	**6**	**7**	**8**
Area (ha)	7929	1496	798	429	290	394	230	427
Stock units	10,113	6639	6310	4068	2865	3727	2647	3547
Stock units/ha	1.3	4.4	7.9	9.5	9.9	9.5	11.5	8.3
Total sheep numbers	8585	4410	4057	2448	1522	2199	2564	1823
Labour units	2.8	2.0	2.0	1.6	1.5	1.6	1.3	2.7
Ewe lambing %	101	122	121	130.9	133.6	133.5	141.3	133
Shorn wool sold (kg per sheep)	4.1	4.2	4.6	5.0	4.2	4.3	5.0	3.0
% of gross income from wool sales	34.5	14.3	10.9	10.2	5.6	9.3	14.8	1.5
% of gross income from sheep sales	35.7	50.0	47.7	46.2	32.0	41.8	67.9	11.3
% of gross income from beef sales	15.5	22.7	36.0	34.5	45.1	20.3	4.1	2.5
% of gross income from other sales	14.3	13.0	5.4	9.1	17.3	28.6	13.2	84.7

Source: Beef & Lamb New Zealand (2016c).

Farm class types: 1 = South Island high country; 2 = South Island hill country; 3 = North Island hard hill country; 4 = North Island hill country; 5 = North Island intensive finishing farms; 6 = South Island finishing-breeding farms; 7 = South Island intensive finishing farms; 8 = South Island mixed cropping and finishing farms.

Merino rams.

Calendar of sheep farming operations

This calendar is provided as a brief guide to the timing of events on a sheep farm, but does not contain detail on optimal management. It is mainly hypothetical because of the wide variety of sheep farming systems. In this scenario, mature ewes are lambing in early September and ewe hoggets (ewe lambs) are bred at eight or nine months of age to lamb in October. In New Zealand only about 30–35 per cent of ewe hoggets are bred, with the majority of animals bred for the first time at 18–19 months of age as a two-tooth.

A general rule of thumb is that lambing begins earlier in the North Island, especially on the east coast, than in the South Island. The latest lambing occurs in the high country of the South Island. The net effect of this is that most lambs are born from early July to late October, with the majority born in late August/September.

This calendar does not give exact feeding guidelines. However, it has been reasonably well established that on a ryegrass/white clover pasture, if pasture heights do not drop below 4 cm and the pasture is of good quality, then intake is not limited.

Once pasture cover drops below 2 cm the sheep will be receiving a maintenance level of nutrition. It is also known that once the pasture height of a ryegrass/white clover pasture rises above 10 cm, especially in summer and autumn, intake may not be restricted but the quality of the herbage will be poor and therefore animal performance will not be optimal.

The growth of newborn lambs is heavily dependent on the amount of milk they obtain from their mother, which in turn depends on the genetic make-up (breed), the number of lambs suckled, the body condition of the ewe and her nutrition. Yields of milk from well-fed ewes peak at 2–5 weeks of lactation and most ewes lose body condition over this period. Liveweight gains of 300 g per day or more in young lambs can be achieved if the ewe is provided with a high allowance of a good-quality pasture.

A ewe rearing twin lambs does not produce twice the milk volume of a singleton-rearing ewe, so the level and quality of feed available is vitally important for ewes rearing multiple lambs. Lambs begin to eat herbage by 3–4 weeks of age. However, while multiple-born lambs may spend more time grazing than their singleton-born counterparts, due to the size of their rumen they generally fail to compensate for a lower individual milk intake. This explains why, on average, twin and triplet lambs are lighter at weaning. Providing high-quality herbage that can include herbal clover swards (e.g. plantain, chicory, lucerne) is one way to increase weight gain.

OCT

Removal of lambs' tails (known as docking or tailing) usually occurs at 3–8 weeks of age. It is done on most farms to reduce the build-up of faecal material (dags) around the breech and reduce the risk of flystrike (blowfly maggot infestation). Ram lambs traditionally were castrated at the same time, but the demand for leaner carcasses has meant that more ram lambs (which are used for meat) are now either left entire or turned into cryptorchids (their testicles are pushed up against the abdominal wall or into the body cavity to render them infertile; another term is short-scrotum). At this time lambs may also be ear-marked or ear-tagged for identification purposes. They may be vaccinated against clostridial diseases (diseases caused by the *Clostridia* bacteria, including pulpy kidney disease, tetanus and black leg) and scabby mouth (orf), be given a trace element supplementation (e.g. vitamin B12) and, in warmer parts of the country, may have their tail area sprayed with an insecticide to prevent flystrike.

NOV

During this period ewes are still lactating. The farmer will be considering ram requirements for the following year and purchasing new rams. Rams will be selected on both genetic merit (as indicated by their breeding values) and physical appearance. Ideally, special attention will be paid to the breeding values of relevant economically important traits, such as fertility (lambing percentage), weaning weight (lamb growth) and potentially fleece characteristics. Rams are normally purchased at 14–16 months of age and used on farm for 2–4 years. Farmers will sometimes have a mix of rams in their ram team: some will be used to breed ewe replacements, while others, termed terminal sires, will be used to breed lambs of which all are destined for slaughter. A third group of rams might be kept specifically for mating with ewe hoggets. The breeds or traits of these rams will ideally result in their lambs being relatively small at birth. Giving birth to large lambs that can cause birthing difficulty (dystocia) can be very detrimental to the success of hogget breeding.

On average, lambs are weaned at 10–14 weeks of age. Under ideal feeding conditions, a high proportion of the male lambs will be sold directly to slaughter as prime lambs (32–38 kg liveweight, 14–16 kg carcass weight). In New Zealand, the average lamb liveweight at weaning is likely to be around 28 kg. At weaning, farmers often decide which ewe lambs will be kept to become replacement ewes; generally, 25–30 per cent will be kept. Lighter ewe lambs are generally not kept as replacements. Farmers have two options for lighter-weight male and ewe lambs — keep them to grow them out to slaughter weight (finishing prime lambs), or sell them to another farmer to finish (store lambs). The decision on how many to keep or sell is dependent on the amount and quality of herbage the farm will grow in the following summer–autumn period. The lambs that are kept may be regularly weighed during the summer and autumn and sold to slaughter when they reach their target slaughter weight.

Finishing prime lambs and replacement ewe lambs will require preferential feeding on good-quality pasture with heights above 4 cm over the summer–autumn period if they are to grow at a rate of 150+ g/day. In New Zealand, the average growth of weaned lambs on pasture is likely to be less than 100 g/day, highlighting the difficulty that many farmers have in achieving high lamb growth rates on ryegrass/clover pastures. The target slaughter weight in the period following weaning is likely to be a minimum of 38 kg, due to lower dressing-out percentages, with carcasses averaging 17–19 kg. The faster a lamb can be grown to slaughter weight after weaning, the more efficient the finishing system is, as less feed is consumed for maintenance requirements. Moreover, the lambs are exposed to fewer potential health problems. Well-grown ewe lambs and hoggets will tend to have a higher lifetime wool and lamb production. To help achieve high lamb growth rates, many farmers use alternative permanent herbages and short-term crops over this period.

Young sheep are particularly susceptible to internal parasites (worms). In conjunction with their veterinarian, farmers should have a parasite control plan in place for the period from weaning to the following winter. This plan generally involves lambs receiving their first de-worming medication (anthelmintic or drench) at weaning, and then being drenched every 28 days for the following four months with potential further drenches if required. Lambs that are going to remain on farm should also be vaccinated at weaning to protect them from clostridial diseases.

After their lambs are weaned, ewes which are deemed no longer suitable as breeding stock are sold on to other farmers or to slaughter (culled). The culling of ewes is usually based on poor reproductive performance (i.e. failure to lamb, failure to rear lambs to weaning, lambing late) or physical condition (i.e. poor teeth, feet or udder).

JAN/FEB

Shearing of the ewe flock is commonly undertaken during the summer period, often coinciding with weaning. Annual ewe fleece weights are generally in the range of 3.0–5.0 kg for Merinos and 4.5–5.5 kg for strong-wool breeds. Lambs are generally shorn after the main flock. One reason to shear ewes and lambs at this time is to help prevent flystrike over the summer–autumn period. Additionally, although there is no clear evidence for this, many farmers believe that shorn lambs grow faster because their intake will not be reduced due to heat stress in hot summer conditions. A few weeks after shearing, most farmers will also dip or spray their sheep with an insecticide at least once to prevent flystrike occurring.

Although at this time the focus is generally on young livestock, the management of ewes between weaning and rebreeding has a significant impact on the number of lambs they will wean the following spring. Ewes that are in good body condition when bred will have higher pregnancy rates, conceive earlier in the breeding period and be more likely to rear multiple lambs than those in poor condition. At weaning, farmers should therefore identify poor-condition ewes that they intend to keep, and manage these to enable them to regain body condition. Those ewes already in good condition can be used to clean up or control poor-quality herbage to ensure that good-quality herbage grows and becomes available for the growing lambs. This type of targeted feeding ensures that those ewes who would most benefit from additional feed receive it, while also ensuring that appropriate levels of feed are available for growing lambs and not wasted on ewes which will not benefit from it.

The amount and quality of semen that a ram produces is affected by his nutrition, body condition and health status. It is therefore important that rams are well fed and in good body condition and health during the eight weeks

Feeding lamb.

leading up to breeding. In addition, rams should be checked by a veterinarian eight weeks before mating to ensure that they are in good condition (sound) and are free from diseases such as brucellosis or scrotal mange. Early checking of rams allows time for any potential issues to be addressed. During the pre-breeding period, farmers may carry out a further examination of ewes to determine whether any more need to be culled prior to breeding.

Ewe ovulation rate is positively related to ewe liveweight and body condition (called the static effect), and also to plane of nutrition (called the flushing or dynamic effect) pre-breeding. The flushing effect is greater in lighter ewes in poor body condition than in heavier, better-conditioned ewes. Farmers therefore flush their ewes in the six-week period before mating. To achieve this, the ewes need to gain weight at a rate of at least 100 g/day, which requires a good-quality ryegrass/white clover sward with a pasture height above 4 cm. As these grazing conditions are generally not available on most farms in autumn, farmers may choose to only flush the lighter, poorer-condition ewes.

On most farms during this period, ewe lambs move from being called a lamb to being called a hogget. If ewe hoggets are to be mated they should be weighed regularly, ideally monthly, to ensure that they are on target to be a minimum of 40 kg at breeding in early May. During the period between weaning and breeding, these ewe hoggets will need to be vaccinated against toxoplasmosis and campylobacteriosis, which cause abortion. On farms where ewe hoggets are not bred, these animals are bred one year later for the first time as two-tooths and vaccinations are generally given the following year.

APR

Ewes will normally come into oestrus (in heat) and be receptive to the ram every 16–17 days between February and late June. Merinos, Dorset Horns and Poll Dorsets have a longer breeding season and can be mated slightly earlier or later; there are currently no breeds in New Zealand that consistently breed all year round. The mature ewe will be in oestrus for 26–36 hours and during this period will be mated on several occasions, often by more than one ram. The length of pregnancy is 142–155 days.

Generally, farmers put teams of rams with the ewes for 34–51 days (2–3 breeding cycles). With ewes in good body condition and with sound rams, pregnancy rates above 95 per cent can be achieved in 34 days, leading some to question the wisdom of a third breeding cycle — this might result in a few late-born lambs, which can be difficult to manage. Often farmers will use a ram of the same breed as the ewes for the first cycle (17 days), to produce replacement ewe lambs. Commonly, 70+ per cent of ewes will be pregnant after the first cycle.

Some farmers may then replace these rams with terminal sires (e.g. South Suffolks, Suffolks, Polled Dorsets or Texels), with the objective of producing prime lambs destined for slaughter. An alternative approach is to select those ewes that the farmer believes are most suitable (e.g. younger ewes, ewes that are high performers) to produce replacements from. All of these ewes will go to a ram of the same breed for the entire mating period, with the remainder of the flock being bred to terminal sires.

During the breeding period, farmers will generally use a ram to ewe ratio of 1:100, although ratios of 1:150 are likely to be just as effective, especially on easier country with smaller paddocks. Farmers may also place crayon harnesses on the rams and may change the colour of the crayon during the breeding period. These harnesses leave a coloured mark on the rump of the ewe during mating, which can be used to identify when she was mated and therefore when she will lamb. This information allows the early culling of non-pregnant ewes after breeding. It can also be used to identify early-lambing or late-lambing ewes, allowing farmers to time when to provide additional feed in late pregnancy and when to set stock for lambing. Ideally, ewe intake would not be restricted during the breeding period.

The first oestrus (puberty) in many ewe hoggets occurs at 7–10 months of age and at about 60–65 per cent of mature weight (38–40 kg liveweight). About 70–90 per cent of the lambs born in the spring will show oestrus during their first autumn/early winter period. The heavier the ewe hoggets are at breeding, the more likely they are to get pregnant and successfully rear a lamb to weaning, and the greater their likely lifetime performance. There are thus clear reasons for farmers to target heavier ewe hoggets at breeding. Some farmers choose to expose ewe hoggets to vasectomised rams (with the vas deferens cut, rendering them infertile) for 17 days pre-breeding. This has been shown to result in more hoggets coming into oestrus during the first 17 days of the subsequent breeding period, and therefore a greater proportion of hoggets becoming pregnant early in the breeding period.

At the end of the mature ewe breeding period, ewes are generally placed on a maintenance level of feeding. To achieve this, the ewes are put on a slow winter rotation around the farm to control intake. This also allows pasture cover to increase during winter, with the aim of ensuring that suitable levels of herbage are available in late pregnancy and at set stocking (sheep placed in a paddock at a defined rate of stock units per hectare). The additional nutritional requirements of pregnancy are low in the first 50 days of gestation. Embryonic loss in the first three weeks of gestation can occur in stressed animals, and stresses such as shearing, transportation and severe underfeeding should therefore be avoided.

The breeding of ewe hoggets traditionally begins in early May. Ewe hoggets are termed shy breeders, as they are less likely to seek the male, are in oestrus for a shorter period and are less likely to stand for the ram than mature ewes. When breeding ewe hoggets, a lower ram to ewe ratio (1:50) is therefore advocated and ram hoggets should not be used unless this ratio is even lower. Farmers also need to consider carefully the rams that they use. While most farmers do not select progeny born to ewe hoggets as replacement animals, there is little evidence to suggest that they shouldn't — and from a genetic gain perspective it would be advantageous to select replacements born to ewe hoggets. When selecting a ram for hogget breeding, farmers should avoid rams that are likely to result in large lambs at birth, as these young ewes, due to their physical size, are very susceptible to birthing difficulties.

Generally, rams are bred with ewe hoggets for 25–34 days, to avoid the impact of very late lambs as the lambing of the ewe hogget flock begins at least a month after that of the mature ewes. As with mature ewes, using crayon harnesses on rams provides advantages for farmers. During the breeding period and in gestation, ewe hoggets should be fed at a level that ensures their continued growth in addition to the nutritional needs of their pregnancy.

Approximately 45–50 days after rams have been removed from the mature ewe flock, ewes should be pregnancy scanned via trans-abdominal ultrasound to identify non-pregnant ewes and the number of fetuses carried by each pregnant ewe. Non-pregnant ewes are culled, being sent either to the sale yards or direct to a slaughter plant. Pregnancy scanning is an ideal time for farmers to assess the body condition score of their ewes. Using this information coupled with pregnancy data, farmers can prioritise multiple-bearing, poor-condition ewes for increased feeding levels in mid-pregnancy.

Pasture growth rates are usually lower than total flock requirements in this period, so most mature ewes are still on a maintenance level of nutrition. Many farmers use winter crops, often based around brassicas such as swedes, turnips and kale, as pregnancy feed for ewes. Use of these generally high-yielding crops allows pasture cover to build up on the farm for late pregnancy and lactation. In parts of the country with very cold winter weather conditions, which have a large impact on pasture growth, ewes are managed behind an electric wire with daily shifts. This ensures very strict control of ewe intake. Ewe

Winter shorn Romney-type ewe with twins.

pasture intake may be supplemented with hay and pasture silage in some farming systems.

In mid-winter, pregnant and even non-pregnant hoggets should be given feeding priority over pregnant mature ewes and should be gaining liveweight (130 g/day and 50 g/day, respectively). Failure to ensure the growth of hoggets during this period can have long-term consequences. Following breeding, rams are placed back on a maintenance diet until eight weeks prior to rebreeding. Older rams which are not going to be used the following autumn are culled.

The nutritional demands of pregnancy increase in mid-pregnancy before rising significantly in the last 50 days of gestation. Using targeted feeding practices, farmers will begin to further increase the feed allowance of multiple-bearing mature ewes, especially those in poor body condition. Increased allowances for singleton-bearing ewes can be delayed until later stages of pregnancy.

Hoggets are pregnancy diagnosed to determine the number of fetuses, and those found to be non-pregnant are generally retained and kept for rebreeding as a two-tooth. On a few farms, however, farmers present more ewe hoggets for breeding than they need for replacements and then sell the non-pregnant hoggets.

During the last three weeks of pregnancy, mature ewes — especially those bearing multiples — should now be offered ad libitum feed (given as much feed as they will consume). Ewes should be moved off winter brassica crops, as these can be bulky and may make it difficult for ewes to consume their theoretical needs. Generally, ewes are allocated to their lambing paddock in the final two weeks prior to the start of lambing. This is termed set stocking; the ewes remain in the same paddock, often through to weaning. Ideally, ewes will be stocked at a rate at which the predicted pasture growth will equal the feed demand of lactation, while ensuring that pasture heights are maintained in the 4–8 cm range. To achieve this, singleton-bearing ewes are set stocked at a higher rate than multiple-bearing ewes. Those ewes that became pregnant later in the breeding period may be kept on a winter rotation a bit longer and then set stocked after the earlier-lambing ewes. This management approach saves feed, but relies on knowledge of the breeding pattern. A further mechanism for efficiently matching feed supply with feed demand is to set stock ewes with a poorer body condition at a lower stocking rate than those in better condition.

Ewes are generally given a booster vaccination against clostridial diseases approximately 2–6 weeks before lambing, to ensure that the lambs receive adequate levels of antibodies via the colostrum. Also a few weeks prior to lambing, any ewes that have not been winter shorn are given a full belly crutch. This involves removing the wool from around the udder and the breech to make those areas cleaner and give lambs better access to the teats. Some farmers may also drench their ewes at this time.

An improved level of feeding in late pregnancy should reduce the possibility of ewe losses occurring from metabolic disorders such as ketosis (twin-lamb disease or sleepy sickness) and hypocalcaemia (milk fever). It will also avoid ewes being weak during birthing, which may result in an extended birth process, birthing difficulties, lighter lamb weights at birth, low lamb energy reserves, impaired lamb thermoregulation, less vigorous lambs and poor mammary gland (udder) development which can reduce colostrum and milk production. All of these can have negative impacts on lamb survival and growth.

In New Zealand, the shepherding of ewes during the lambing period varies from intense, regular observation of ewes and lambs to much less or no oversight (often called easy-care). On large farms where human–livestock interaction can be minimal and often associated with stressful events such as shearing, minimising interference in the ewe–lamb bonding process during and after birth can decrease lamb mortality through mis-mothering.

However, there are some advantages to closely monitoring ewes at lambing: lifting cast ewes (those that are stuck on their back) back on to their feet, attending to bearings (vaginal prolapse), assisting with lambing difficulties and mothering-up orphan lambs. In addition, many ram breeders collect dam–offspring information. If lambing management is intensive, it is important that the ewes are accustomed to human intervention as otherwise good intentions can have adverse effects.

Pregnant hoggets should continue to be fed at a level that allows them to gain at least 130 g/day. As the hoggets move into the latter stages of pregnancy, more of this gain is associated with the pregnancy. Many farmers make the mistake of controlling feed intake in the later stages of pregnancy of hoggets, with the aim of avoiding large lambs and associated dystocia. However, it is the failure to grow out hoggets in early pregnancy that causes dystocia, as a small-framed hogget is more likely to have difficulty giving birth due to a small pelvic opening. As with mature ewes, hoggets should be given a belly crutch, appropriate vaccinations and possibly a drench prior to lambing.

Merino ram and ewes.

South Island high-country farms

As outlined earlier, there are eight sheep farm classes, and therefore actual management on any individual farm will vary accordingly. For example, South Island high-country farm management differs significantly from that in the other sheep farm classes. There are about 200 South Island high-country farms, also called stations or runs. These are usually based at the higher altitudes in the ranges of Marlborough, Canterbury and Otago. Individual farms often occupy large areas of land, much of which is steeply contoured and covered by native tussock. On these farms, Merinos, New Zealand Halfbreds and Corriedales are the predominant breeds, followed by Perendales or composites based on these breeds. Some have a large number of wethers (castrated males), solely for wool production, usually on those areas that are more difficult to access. Due to the environment and the breed types farmed, lambing percentages tend to be lower than for other farm classes. The management of these high-country farms is often vastly different from other sheep farms, being governed by the significant snowfalls that tend to occur over much of the grazing land from about May through to October. Sheep must therefore be restricted to the lower parts of the farm for much of this period. A typical farming calendar might include an autumn muster to collect ewes from the snow-prone higher areas and move them down towards the wintering blocks within the farm system, which are at a lower altitude. The rams are run with the ewes on these wintering blocks, usually from early May. Then, from mid to late spring onward, the sheep are gradually moved back onto the high summer grazing areas. Lambing starts in October, with shearing usually undertaken before the ewes move towards the more remote parts of the farm.

Shearing

On most farms it is now standard practice to shear ewes once yearly, although if the breed allows for it then some farmers will shear twice yearly. Merinos and many terminal breeds do not grow a long enough fleece to allow for twice-yearly shearing. In contrast, the Drysdale's annual fleece growth is such that twice-yearly shearing is required. There are minimum and maximum fibre lengths for processing that influence the shearing policy for a given breed.

The reasons why some farmers do not shear twice yearly include the value of the wool being outweighed by the cost of shearing; difficulty in ensuring that ewes are dry for shearing; increased risk of ewe deaths if shearing in cold weather; and increased feed demand following shearing. Conversely, some reasons for shearing twice yearly include improved lambing percentages; faster lamb growth rates; fewer cast sheep; increased wool production; better wool colour; less wool in the oddment categories; less crutching and dagging; and a more even cash flow. There are other advantages and disadvantages, but their scientific validity has not been proven, and indeed some of those mentioned above depend on feeding conditions, ewe body condition and the timing of shearing during pregnancy.

A few farmers have a split-flock shearing policy, with three shearing times a year but only half of the sheep being shorn at each shearing time. This results in each sheep being shorn at intervals of approximately eight months. Each shearing mob contains ewes from different age groups. For example, a ewe could be shorn as a lamb in January 2016 (at four months of age), as a hogget in October 2016 (at 14 months of age), then in April–May 2017, January 2018, October 2018, etc. Every January, therefore, there will be lamb's wool harvested in addition to wool from the two- and four-year-old ewes. In May, wool will be harvested from one- and three-year-old ewes and in October from the hoggets and the 2.5- and five-year-old ewes.

Shearing time

Shearing time affects wool quality. With twice-yearly shearing, the characteristics of each of the two clips in the year differ: wool shorn in March will differ from that shorn in late October/early November. Wool shorn in June (mid-pregnancy) differs from that shorn in late December/early January. The discussion here on the impact of shearing time on wool characteristics will be restricted to a once-yearly shearing policy, as this is currently the most common policy on New Zealand farms. In addition, unless otherwise stated the effects outlined below are for Romney-type wool (32+ microns) from a mature ewe lambing in September.

During October/November, the weather becomes warmer and the risk from cold stress declines. At this time the wool is generally of good colour, and any tender regions in the fibres arising from a winter weakness are close to the base of the fibres. Cotts (extreme fibre matting in the fleece) tend to be minor if present at all. Plants have not yet gone to seed, so seed and plant material contamination tends to be low. A disadvantage of shearing at this time, and probably the main reason why farmers do not shear then, is that the lambs are still with the ewes. The weather also tends to be unsettled during this period and ensuring that sheep are dry for shearing is not always easy.

Shearing during the December–January period is often the most convenient time for the farmer, although it is not ideal from a wool quality perspective. In this period there is only a very low risk of death from cold stress. Shearing is generally timed to occur at or near weaning, and both ewes and lambs can be shorn. Shearing in this period provides a level of protection from flystrike during a time of risk. However, by December considerable levels of wool yellowness will begin to occur due to humid weather. Cotts become increasingly entangled and the tender region of the wool, if present, will be towards the middle of the fibres. Many plants have gone to seed by now, so contamination of the wool can become an issue.

Shearing in the April–May period is probably done as part of a twice-yearly (second shearing) policy. Once-yearly shorn wool at this time has the tender region near the tip of the fibres. Colour and vegetable-matter contamination of the wool are generally not too bad at this time, due to the length of the fleece during the spring–summer period. Farmers do not generally shear ewes at this time, as it can interfere with breeding and in some areas cooler weather conditions can increase the risk of cold stress.

Shearing in June and July, sometimes called mid-pregnancy shearing, is most often used with a twice-yearly shearing policy. Shearing at this time has been shown to increase lamb survival and potential growth prior to weaning. On high-country stations (predominantly with Merinos, in August/September), ewes are often shorn within the last 30–40 days of pregnancy (late-pregnancy shearing). An advantage of shearing at this time is that the tender region of the fibres is usually close to the end and the fibres tend to have higher tensile strength. Additionally, the colour tends to be good due to the relatively short length of the wool over the spring–summer period. The disadvantages of shearing at this time is that it can put extra pressure on feed demand in winter, and there is a high risk of cold stress and ewe deaths if appropriate measures are not put in place.

As mentioned in the calendar section, ewes that are not shorn in mid or late pregnancy should be given a full belly and breech crutch prior to lambing. Over the summer–autumn period, sheep with dags are very susceptible to flystrike so it is good practice to crutch the breech of sheep, if they have not been recently shorn, to remove the dags and reduce flystrike risk. Abattoirs will not accept animals with dags, due to potential carcass contamination, so farmers need to ensure that any animal with dags is crutched prior to being sent for slaughter. If not undertaken on farm, the abattoirs will do this and charge the farmer accordingly.

Management prior to and during shearing

To minimise the problems that may occur at shearing and to maximise the value of the fleece, appropriate management of sheep prior to shearing is required.

Ideally, sheep approaching shearing should not be kept in paddocks with a lot of seeding plants and dead plant material. A paddock close to the woolshed should be eaten out so that sheep about to be shorn can be placed in it to allow them to 'empty out' before they enter the woolshed. There is also a short period during which the sheep should not have access to water. These measures help minimise the penstain (dung stain) that can occur when sheep defecate and urinate against each other or wipe their dirty rumps on each other.

A non-pregnant sheep should display few ill-effects from being off feed for up to 24 hours before shearing; however, this period needs to be much shorter for pregnant ewes, especially those in the later stages of pregnancy. If sheep are dirty, they should be dagged ideally at least two weeks before shearing to help minimise penstain, as any contamination that occurs then has time to be washed out and/or fade. Sheep of different wool types and lengths generally need to be separated into different lines prior to shearing, as their wool may have different end uses and values. A significant variation in wool characteristics can also impede processing efficiency.

The woolshed should be clean and uncluttered, to reduce potential contamination of the fleece. Farmers should ensure that they have enough staff and ideally enough bins (bales/fages) in which to put the various parts of the fleece (i.e. crutchings, second pieces, wigs, bellies, main-line wool, etc.), to maximise the value of the entire clip; almost invariably, the handling will involve some separation of different types of wool. Much of this separation takes place on the shearing board (often on raised boards), and partly at the wool table.

Belly wools are normally kept separate from fleece wools, as they tend to be more discoloured and have shorter staple lengths even though actual fibre length may not be that different. Neck wool is often removed, especially if it is matted and/or has seed contamination. Cotted fleeces should also be removed from the main line and binned separately.

The term skirting refers to the removal of most of the contaminated or faulty parts of the fleece, generally at its edges. From an economic perspective, skirting is really only required if the remaining parts of the fleece are of good colour and the contaminated or faulty wool comprise no more than 10 per cent of the total fleece. With second-shear (six-month wool growth that does not allow the fleece to be collected as a whole), the removal of the shorter, more yellow staples is best carried out on the shearing board rather than on a wool table. If skirting is not carried out, there still needs to be some level of inspection of each fleece as it comes off the sheep, so that foreign matter, dags, serious stains, etc., may be removed.

Preventing cold stress and death from shearing

Shearing in cold conditions, especially coupled with wind and rain, can result in cold stress and death. Sheep have not evolved to lose their fleece suddenly, as occurs with shearing. They are very poor at restricting blood flow to the extremities soon after shearing, and this can take a few days to fully occur.

Certain management steps can be put in place to limit the potential for death following shearing: being aware of the weather forecast and not shearing in cold, wet and windy conditions; using blades or cover (winter) combs that leave a greater stubble depth after shearing than standard shearing combs; limiting the period that sheep are kept off feed prior to shearing; having sheltered paddocks with good pasture cover to ensure adequate nutrition for sheep after shearing; stopping shearing before it gets dark to encourage sheep to graze; and being prepared to return sheep to a woolshed if needed to provide further shelter.

Nutrition

Monitoring nutritional status — liveweight

Liveweight (LW) is an easily obtainable and non-invasive measurement of sheep. Considerable variation exists in the expected mature liveweights of sheep, ranging from 50 kg in smaller hill ewes to 90 kg in lowland ewes. Liveweights may typically be characterised within breeds, given as a standard reference weight (SRW) as an approximation of frame size, but even then the variation in the liveweights of individuals within a breed is frequently greater than the variation between breeds.

Total liveweight can be expected to fluctuate throughout the year; an expected profile is given in Figure 3.3. The dominant factors that influence the change in liveweight throughout the year are changes in nutritional state as a result of deposition or mobilisation of body tissues for lactation, and physiological changes associated with fetus weight during pregnancy.

Long-term fluctuations in liveweight can be expected to reflect changes in tissue reserves. Body protein and bone reserves are generally highly conserved, although some breakdown of lean tissue and utilisation of bone may be experienced during times of nutritional stress and may account for 30 per cent of the change in liveweight. Body adipose (fat) reserves are more variable and primarily account for the remaining 70 per cent variation in liveweight.

In sheep, as in all ruminants, the measurement of liveweight is affected by the weight of undigested material in the rumen (gut-fill), which may be as much as 20 per cent of the recorded liveweight. Gut-fill can change dramat-

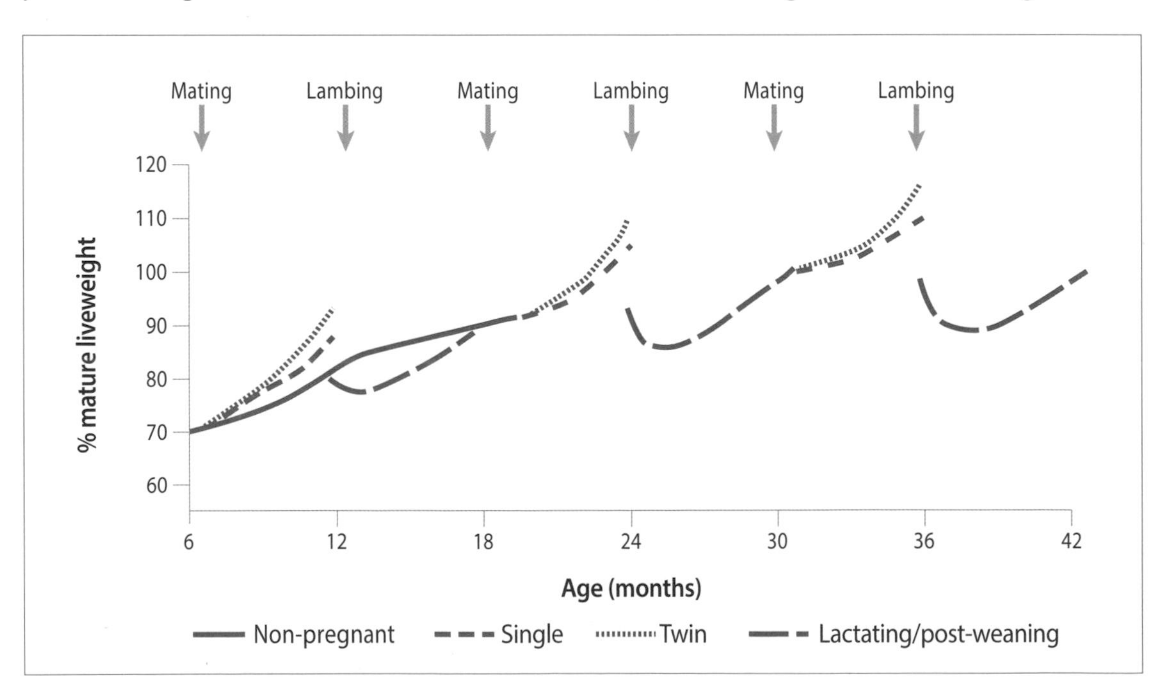

Figure 3.3 Ewe liveweight profile over time.

How to BCS

BCS is assessed by palpation of the lumbar region, on and around the backbone in the loin area immediately behind the last rib and above the kidneys, to examine the degree of sharpness or roundness. The aim is to try to feel the backbone with the thumb and the end of the short ribs with the finger tips. It is a means of subjectively assessing the degree or fatness or condition of a live animal.

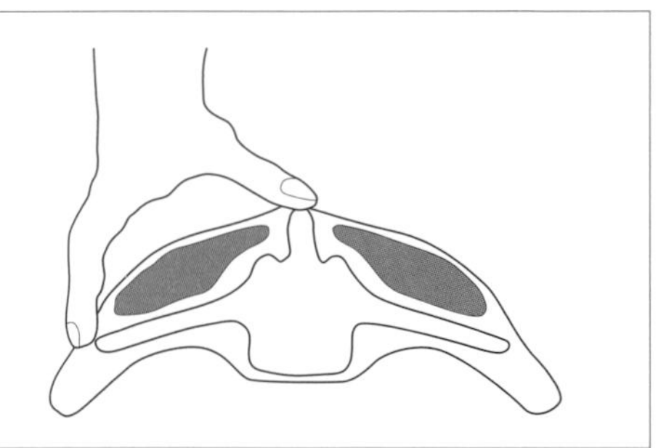

Grade

Score 2 The spinous processes are smooth but still prominent. The individual processes can still be felt, but only as fine corrugations. The transverse processes are smooth and rounded. However, it is still possible to pass the fingers under the ends of the processes with some pressure. The eye muscle area is of moderate depth but has sparse fat cover.

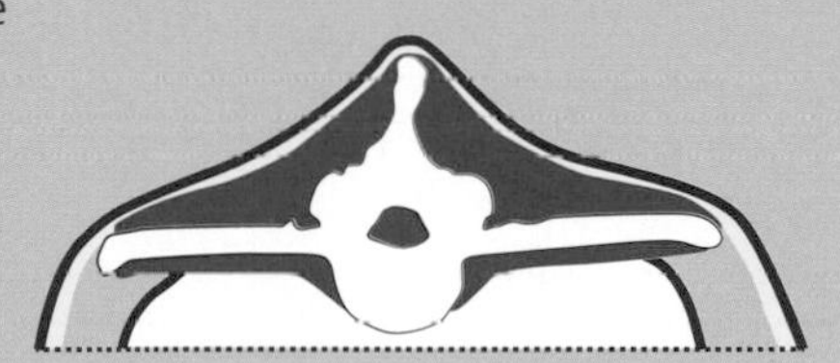

Score 3 The spinous processes are smooth and rounded, and individual bones can only be felt with some pressure applied. The transverse processes are also smooth and are well covered. Firm pressure is required to feel over the ends. The eye muscle area is full and is covered by a moderate degree of fat.

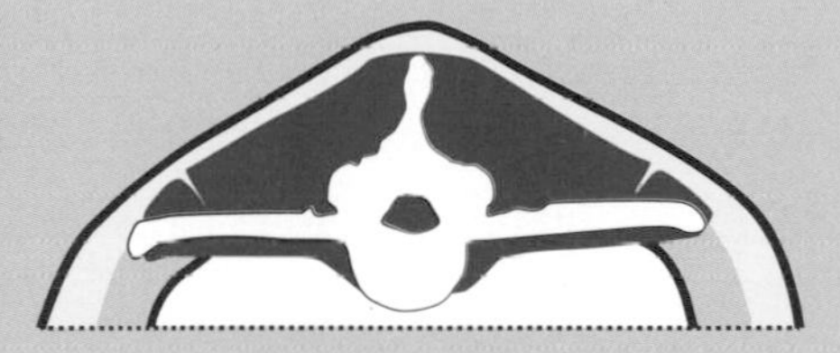

Score 4 With pressure applied, the spinous processes can just be detected while the ends of the transverse processes cannot. The eye muscle area is full, with a thick covering of fat.

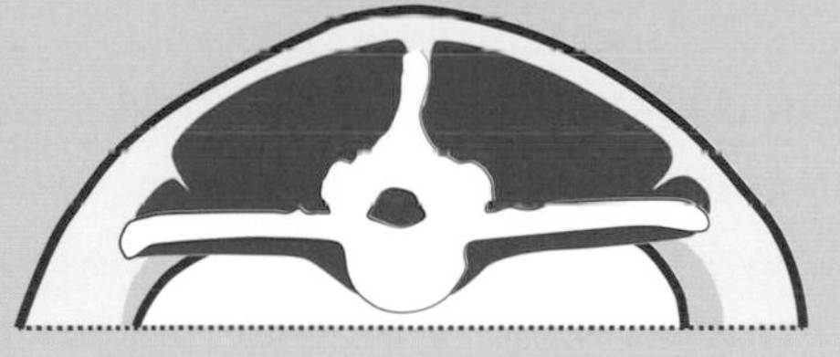

Figure 3.4 Illustration and examples of BCS technique (note that the full range is from 1 to 5, in 0.5-unit increments. Sources: adapted from West et al. (2009) and Kenyon et al. (2014).

ically in a relatively short time, reflecting diurnal variations in grazing behaviour, and depends on the rate of degradation of material through the rumen. Minimum pre-measurement fasting times of six hours have been suggested as a means of significantly reducing the variation in liveweight measurements, although even after 24 hours of fasting the material in the rumen may still be expected to contribute up to 9 per cent of the recorded liveweight.

Monitoring nutritional status — body condition score

Body condition score (BCS) is an assessment of the depth of soft tissue covering the lumbar vertebrae. It is assessed by palpation of the lumbar region and is recorded on a scale of 1–5, often in increments of 0.5, with a BCS of 5 representing an obese animal. Examples of individual BCS values are given in Figure 3.4. The measurement is relatively easily obtainable and provides an effective means of comparing the nutritional status of sheep regardless of frame size. Although it is a subjective measure, repeatability both between and within trained assessors has been shown to be high.

As with change in liveweight, a majority of the change in BCS reflects changes in adipose content and changes in liveweight are thus commensurate with changes in BCS. Published values for the change in liveweight that is associated with a one-unit change in BCS vary from 4 kg to 11 kg, and are largely influenced by animal frame size. In accounting for variation in frame size, a one-unit change in BCS is equivalent to 15 per cent of SRW.

Animal demand

Detailed information on animal requirements for both energy and protein can be found in Rattray et al. (2007) and therefore general guidelines only are given here. Both of these requirements are dependent on physiological state and expected or desired level of performance. Although adequate protein is essential for all aspects of production, energy is generally considered to be the limiting nutrient due to the poor ability of the rumen to effectively process high-quality protein sources.

Maintenance energy requirements are relative to metabolic liveweight ($LW^{0.75}$), being 0.5 megajoules of metabolisable energy per kilogram of $LW^{0.75}$ (MJ ME/kg $LW^{0.75}$) for sheep, with additional energy requirements being added to maintenance for weight change, pregnancy and lactation. The metabolisable protein requirement for maintenance is also dependent on liveweight, being 60 grams of metabolisable protein (g MP) per day for a 60 kg ewe and increasing or decreasing by 0.8 g/day/kg for liveweights either above or below 60 kg.

Feed quality and intake constraints

The ability of sheep to consume forages is physically limited by feed quality. When consuming highly digestible feeds of around 85 per cent digestibility or greater, maximum intake can be as high as 175 grams of dry matter per kilogram of $LW^{0.75}$ (g DM/kg $LW^{0.75}$). With lower-quality feeds, greater proportions of structural carbohydrates increase the time taken for material to degrade in the rumen, resulting in maximum intakes being reduced to around 80 g DM/kg $LW^{0.75}$ with this diet supplying 9 MJ ME/kg DM. In practice, ruminants rarely achieve their maximum potential intake when grazing, as intake is also constrained by herbage availability, grazing time, disease status, presence of a fetus or fetuses and signals from high-quality diets affecting intake. Assuming that sufficient pasture masses, in excess of 1200 kg DM/ha, are supplied, as a general rule of thumb the maximum daily DM intake of a non-pregnant disease-free animal can be estimated as 3.5 per cent of LW.

Teeth

At any time in the production cycle, tooth condition is an important determinant of the ability of a sheep to consume sufficient nutrients. Due to the fibrous nature of forages and the abrasive properties of soil, which may be ingested during hard grazing or when consuming winter forages, tooth wear is both common and expected. Culling decisions based on animal age are frequently associated with teeth that are in poor condition.

Shorn ewes.

Seasonal feed demand

Breeding ewes

For a breeding ewe, energy and protein requirements are relatively low prior to breeding, being predominantly those required for maintenance. Before rams are introduced, additional flushing feed may be given to increase ewe liveweight and take advantage of the known static and dynamic effects of ewe liveweight on ovulation rate. Ovulation responses are dependent on ewe body condition, being greater in ewes with a BCS of less than 2.5 and negligible in ewes with a starting BCS of greater than 3.5. This is also reflected in changes in liveweight, with a 2 per cent increase in ovulation rate being observed for every 1 kg increase in liveweight at the time of ram introduction (joining weight), with little apparent benefit to ovulation if a ewe's joining weight exceeds 68 kg (although breed differences may exist, depending on frame size). Flushing feed requirements can typically be met through the availability of pasture, although in situations where autumn rains have not occurred to stimulate pasture growth, supplementation with energy-dense feeds, such as grains or grain-based pellets, may occur.

Following mating, ewe feed demands are typically maintenance level for the first half of pregnancy, during which time the use of winter forage crops is common. During late pregnancy (following pregnancy scanning), ewe nutrient demand increases due to the increasing demands of the fetus. For an expected lamb birth weight of 4 kg, ewe energy and protein demands at six weeks before lambing may be increased by 2 MJ ME/day and 11 g MP/day per fetus. These values increase to 6 MJ ME/day and 33 g MP/day by birthing time. Immediately prior to birthing, a twin-bearing 60 kg ewe may thus require 22 MJ ME/day and 126 g MP/day. The situation changes dramatically following lambing, when the demands of lactation increase both the energy and the protein requirements in excess of what can be supplied from pasture. Mobilisation of ewe lipid reserves to overcome energy deficits is common, resulting in a decrease in ewe BCS. However, the metabolisable protein requirements are so high that they still cannot be met through this process, resulting in the relaxation of immunity to gastrointestinal parasites immediately before and after birthing as nutrients are diverted from the protein-demanding immune response into milk production.

Supplementation studies have shown that an MP supply of more than 300 g MP/day is required to prevent this relaxation in immunity and fully meet the MP requirements of the ewe; an amount which cannot physically be obtained from consuming pasture alone. Supplementing ewes on pasture during this time with high-quality protein feeds, although theoretically possible, has practical and financial limitations. Post-weaning ewe requirements typically drop to maintenance levels plus, as required, replacement of mobilised tissue reserves prior to rebreeding. Pasture, where available, is typically sufficient to supply both energy and protein for this purpose.

Lambs

The demand for energy and protein for lamb growth is primarily governed by the composition of gain, this itself being a function of

A heavily pregnant ewe.

relative maturity. With increasing maturity, the relative rate of protein deposition decreases and commensurate increases in adipose deposition occur. For the very young lamb, the requirement for protein relative to energy is high, of the order of 11 g MP for every 1 MJ ME; this decreases to around 6 g MP per MJ ME by the time a lamb reaches 30 kg. This high protein requirement highlights the value of milk as a feed source, as this may supply 12 g MP per MJ ME compared with the 6 g MP per MJ ME typically supplied by pasture. Weaning a lamb at less than 30 kg may therefore be expected to cause a protein deficiency and restrict potential growth if suitable nutrition is not given.

Following weaning, once a lamb is above 30 kg, lamb growth is primarily governed by an adequate supply of ME rather than MP. The ME requirements for growth vary according to the sex and the relative maturity of the animal, being approximately 4.5 MJ per 100 g gain for a 30 kg female with a mature body weight of 60 kg, and 3.25 MJ per 100 g for a male of the same weight with a mature body weight of 100 kg. The feed requirements for the maintenance and growth of a 30 kg female growing at 200 g/day are therefore around 14.5 MJ ME/day. Due to the physical constraints of feed consumption mentioned above, providing high-quality diets is critical to facilitate high lamb growth rates.

Feed sources

Pasture as a feed source

New Zealand sheep production primarily relies on ryegrass/white clover pastures as the predominant feed source. For the majority of situations, this combination of grass and legume provides an adequate nutrient supply (although some notable exceptions exist). From a nutrient supply perspective, the energy supply of these pastures is typically within the range of 10–11 MJ ME/kg DM, although values of 11.5 MJ ME/kg DM may be encountered in lush growth and values as low as 9 MJ ME/kg DM once plants have started seeding, which may be expected to occur in late summer, particularly where moisture stress is present. From a protein perspective, the crude protein (CP) percentage of ryegrass/white clover pasture is typically 18–20 per cent or greater, varying with the level of clover, and is generally adequate for synthesis of microbial protein (for which a minimum of 16 per cent CP is typically required). For the majority of situations, therefore, around 80 g MP is supplied per kg DM consumed.

Matching feed supply and animal demand

The seasonal calendar of events in sheep production typically follows the pattern of pasture production, namely the matching of peak pasture growth in spring with peak animal demand during lactation. There are, of course, some considerable regional variations in the ability of pasture to meet animal demand.

Two main factors influence pasture production in New Zealand — temperature and moisture — and the adequacy of these for pasture production in any given year or region is difficult to predict. Where there is adequate moisture, pasture production is generally limited by temperature, particularly during the winter period when pasture production in the south and at higher altitude is negligible and results in a mismatch between feed supply and animal demand. Moisture also demonstrates considerable regional variation, with its distribution being largely influenced by mountain ranges. During summer the east coast of both main islands is prone to inadequate rainfall and high evaporative losses. Relatively few sheep are run under irrigation in these areas. The summer period is a time when animal feed demand may not be met through pasture, necessitating the use of supplementary feeds and/or specialist forage crops.

Ewe on mixed plantain sward.

Winter feeding and forage crops

Although ewe feed requirements (in early pregnancy) are low during the winter period, in a majority of New Zealand environments the total feed demand exceeds that available from pasture growth. Supplementary feeding in the form of conserved pasture (hay or silage/baleage) is common. This may be fed to animals either strip-grazing pasture, particularly in more intensive lowland systems, or being fed winter forage crops. The energy content of conserved pasture is generally lower than pasture, with 1–2 MJ ME/kg DM being lost during conservation, making it difficult for animals to gain weight and/or body condition when conserved pasture constitutes a large proportion of the diet. The primary function of conserved pasture supplements at this time is thus to provide gut-fill to maintain animal satiety.

Winter forage crops are used to transfer nutrients grown during the summer and autumn periods into the winter period to fill the feed deficit left by insufficient pasture growth. Typically these crops are a brassica species, although not exclusively. Swedes and kale are common in Otago and Southland, whereas in warmer and drier environments kale and rape may be used. Fodder beet is an increasingly common winter forage crop that is grown throughout the country.

Winter forage crops are planted in the late spring/early summer and grow during summer and autumn, providing a standing feed bank which is grazed in situ. Depending on environment and forage type, such crops may provide anywhere between 5000 kg DM and 30,000 kg DM per hectare. Most brassica species will supply around 10–11 MJ ME/kg DM; fodder beet tends to be more energy dense, supplying 12–13 MJ ME/kg DM. Crude protein levels tend to be below what is required for optimal rumen microbial protein synthesis, being 11 per cent in swedes, 16 per cent in kale and 8 per cent in fodder beet.

Although the crops are highly digestible, maximum potential feed intakes are seldom observed. This is partly because the crops have a high water content, but is mostly due to grazing management whereby rations are restricted to match a maintenance level. Grazing generally involves large numbers of animals on a relatively small area, resulting in high effective stocking rates which negatively influence both soil properties and nutrient leaching and lead to possible environmental effects that may need to be considered. Commonly, forage crops are strip-grazed with a fresh allocation provided to the animals at regular intervals; these range from daily to weekly shifts depending on farm management.

It is important to adjust sheep to eating forage crops. Brassicas contain secondary compounds such as S-methylcysteine sulphoxide, glucosinolates and nitrates, the levels of which depend on the growing conditions. Time is required for rumen microbes to adjust to, and subsequently metabolise, these compounds. Adjustment consists of a gradual increase in periods of grazing on forage crops with a return to pasture or supplementation with hay or silage/baleage in between.

Due to the high soluble-sugar content of fodder beet, adaptation is essential to avoid ruminal acidosis. Supplementation with hay or silage that is high in crude protein, such as a legume hay or silage, during grazing of fodder beet has the added benefit of overcoming the deficiency of available protein for effective microbial synthesis.

Lambs on plantain — soon to finish for slaughter.

Summer forage crops/specialist finishing forages

Summer forage crops are a means of providing nutrition at a time of year when pasture mass or quality is insufficient. They are commonly used in areas that are prone to drought conditions, but also in areas where summer rainfall may be reliable but the quality of permanent pastures decreases due to plants seeding. As ewe requirements are generally low during summer, the primary purpose of these crops is the supply of high-quality feed, in terms of both energy and protein, for finishing lambs.

Summer forage may comprise short-term crops planted in late spring, such as annual (Italian) ryegrass or the brassicas, leafy turnip (Pasja) or rape, to provide a single grazing with possible repeated grazing of regrowth. Alternatively, these may be longer-term forages such as lucerne or herb mixes of plantain, chicory and red and white clovers, which may last for several years. Rotational grazing is required to allow plants to regenerate root reserves, facilitate growth and maximise intake. Pure legume forages (e.g. lucerne) or herb/clover mixes provide excellent feed sources for young lambs, particularly where plants containing condensed tannins (e.g. plantain) are combined with legumes, which protects the plant proteins from degradation in the rumen. Chicory contains the secondary compounds sesquiterpene lactones, which have been reported to have anthelmintic effects against some gastrointestinal parasite species.

It is recommended that animals are introduced to new feeds slowly, to allow time for the rumen microbes to adjust to the feed type and any secondary compounds that may exist. Particular emphasis should be given to forage brassicas, where high nitrate levels can cause severe animal health issues in unadjusted lambs. Slow introduction of new feeds is also important to minimise production loss through a lack of recognition of feed types when lambs may be encountering feeds for the first time.

Disease

Parasites

The major diseases of sheep in New Zealand are internal and external parasites, trace-element deficiencies, facial eczema and some infectious diseases. Other important conditions affecting sheep health include lamb mortality, bearings (vaginal prolapse) and dental disorders. Key animal health interventions have been included in the farming calendar section earlier in this chapter, and further information on sheep diseases can be found in West et al. (2009) and general guidelines only are given here.

Gastrointestinal nematode parasites (internal parasites or worms) are the main production-limiting disease affecting sheep. Lambs are most affected due to their inadequate immune system; clinical signs primarily include diarrhoea and weight loss, with possible death if untreated. Ewes may sometimes be affected if they are poorly fed or have other health issues. Internal parasites live in the sheep's gut and produce eggs that are shed in the faeces. These develop into larvae on the pasture, and are then ingested by the sheep and a new generation begins. All sheep farms are affected, and control programmes are necessary to limit the effects. Usually these involve a combination of grazing management to limit the exposure of lambs to larvae, and drenches (anthelmintics or wormers) to kill adult worms and reduce egg shedding in the faeces. Lambs are generally drenched at 28-day intervals from weaning to about eight months of age, but ideally individualised parasite control programmes would be developed for each farm in conjunction with a veterinarian or advisor. Development of drug resistance to anthelmintics is an increasing problem in the control of internal parasites.

Flies are the most important external parasite of sheep, causing the condition called flystrike. Flies are prevalent in warm, humid conditions from October to May and are more likely to lay eggs in malodorous areas of the sheep such as dags or wounds, although some species lay eggs on the rump or shoulder of the sheep without other predisposing factors. The developing larvae (maggots) feed on the sheep's skin and flesh and, if untreated, this results in death. Keeping sheep free from dags will reduce the risk of flystrike, while shearing prevents it for 4–6 weeks. Control is primarily achieved by topical application of insecticides that prevent flystrike for 2–16 weeks depending on the type of insecticide and the method of application.

Nutritional deficiency

Most sheep in New Zealand receive their nutrition from grazing pasture or forage, meaning that they are dependent on the plants, and therefore the soil, for the provision of all nutrients. Some trace elements, which are required in small quantities and are essential for normal cell processes, are deficient in some New Zealand soils. The most important deficiencies in sheep are those of selenium, cobalt and iodine. Selenium and cobalt deficiencies most commonly result in poor growth of young sheep, while iodine deficiency may result in goitre or increased deaths of newborn lambs.

Fungi

Facial eczema is a fungal disease that occurs

during late summer and autumn in the warmer regions of New Zealand. The fungus lives on grass; ingestion of the spores results in liver damage that limits the sheep's ability to detoxify certain compounds found in grass. Accumulation of these toxic compounds leads to photosensitivity. Clinical signs initially include irritation, shade-seeking and swelling of the non-wooled areas (most noticeably the face and ears). This progresses to weight loss and peeling of the skin on the face and ears. A number of other sheep without these skin lesions are likely to also have liver damage.

Facial eczema is a very painful and costly disease. Prevention in sheep can be challenging; options include identifying risk periods through spore counting, grazing management, fungicide spraying of pasture, orally dosing the sheep with protective zinc, and breeding for genetically tolerant sheep.

Infections

Important infectious diseases are those causing abortion (the infecting organisms include toxoplasma, campylobacter and salmonella), and those caused by clostridial bacteria which usually result in rapid death. Lameness due to bacterial footrot can be a problem, particularly in fine-wool (e.g. Merino) flocks. Vaccines are available against a number of infectious diseases; it is highly recommended that sheep are vaccinated against toxoplasma, campylobacter and clostridial diseases.

Vaccines are also available for footrot, salmonellosis (which causes abortion or enteric disease), orf/scabby mouth (skin lesions on lambs, which are transferrable to humans), Johne's disease (causing weight loss and death of adult sheep), leptospirosis (transferable to humans) and caseous lymphadenitis (infection of lymph nodes). These vaccines are generally only used when there is an issue on farm, increased risk of infection or a specific perceived need.

Other issues

Lamb mortality can be a problem in New Zealand sheep flocks, particularly when inclement weather occurs during lambing. The majority of lamb deaths are caused by mis-mothering/starvation, exposure and dystocia. Triplet lambs are particularly prone to mortality due to their lower birth weights and the difficulty that ewes have in mothering and feeding three lambs. Lambing dates vary depending on farm location, to match the spring grass growth and also to increase the likelihood of clement weather during lambing. However, the nature of New Zealand weather means that storms may occur at any time.

Bearings (prolapsed vaginas) most commonly occur in multiple-bearing ewes in late pregnancy, but occasionally occur at other times of the year. The risk factors for development of bearings are largely unknown, but the cause is likely to be multifactorial. Affected ewes require prompt treatment or euthanasia.

Dental disorders due to excessive wear of the incisors, or incisor loss, are common on some farms and often result in ewes being culled prematurely before the end of their potential productive lifespan. Poor teeth result in ewes being less able to feed, leading to weight loss. The causes are largely unknown.

Bottle-reared orphan lamb.

Environmental considerations

The environmental issues faced by the sheep industry revolve around water quality and supply, climate change and greenhouse gas emissions (Morris & Kenyon, 2014). Minimising environmental impacts while achieving improvements in animal performance and economic returns will be a future focus for the industry. As indicated earlier, an increasingly large proportion of the national flock is farmed on hill or high country and there is sparse information on the environmental impacts of farming in these environments or on how to mitigate against potential impacts. Encouragingly, Mackay et al. (2011), using the computer decision support model OVERSEER, found that for hard hill-country sheep and beef farms the productivity gains (a 47 per cent increase in saleable product/ha) achieved since 1989/90 also translated into environmental gains, including a 21 per cent reduction in nitrate leaching per kilogram of saleable product and a 40 per cent reduction in greenhouse gas emissions per kilogram of saleable product.

Animal welfare

Sheep are particularly vulnerable to stress during the loading and unloading processes of transportation. This can be reduced by having well-designed yards, avoiding an overly steep ramp angle, ensuring that the light differential between inside and outside the vehicle is minimal, and using skilled stockpersons (Dwyer, 2008). Time spent in the yards should be minimised and sheep should not be overcrowded in pens. Prior to transport, all animals should be examined and only those that are fit and healthy (able to stand and bear weight on all four limbs) should be transported (New Zealand Government, 2010).

Internationally, the pressure on the World Trade Organization (WTO) to consider the inclusion of animal welfare standards in trade agreements will increase (Ferguson et al., 2014). As an export-focused industry, the New Zealand sheep industry will need to ensure that it continues to place emphasis on sustainable and ethical animal production systems. By continuing to develop and maintain animal ethical and welfare practices that meet or exceed international standards, the New Zealand sheep industry will, in addition to its clean green reputation, be able to further differentiate its products as high value to international consumers that are becoming ever more discerning.

Consumers are also increasingly requiring assurance that the products they consume are safe, not just of high quality. The industry therefore needs to avoid potential issues by showing that it has appropriate animal health, zoonosis (diseases which affect both animals and humans) and food safety/hygiene protocols in place. Full traceability of animals back to an individual flock is likely to be needed, which may require a national animal identification scheme such as that currently in place for cattle. In addition, the sheep industry may need to increase its efforts on educational campaigns to build consumer trust. This will require a transparent and open farm business situation that demonstrates aligned values around the care of farm animals (Ferguson et al., 2014).

References

Anon. 2013. A yarn of wool. *New Zealand Economics ANZ Agri Focus*, April. Wellington: ANZ Research, ANZ Bank New Zealand, 2.

Beef & Lamb New Zealand. (2015). Stock Number Survey as at 30 June 2015. Wellington: Beef & Lamb New Zealand Economic Service. Accessed at www.beeflambnz.com/Documents/Information/Stock%20number%20survey.pdf.

Beef & Lamb New Zealand. (2016a). Compendium of New Zealand farm facts, 40th edition. Wellington: Beef & Lamb New Zealand Economic Service. Accessed at www.beeflambnz.com/Documents/Information/nz-farm-facts-compendium-2016%20Web.pdf.

Beef & Lamb New Zealand. (2016b). Farm classes. Wellington: Beef & Lamb New Zealand Economic Service. Accessed at www.beeflambnz.com/

information/on-farm-data-and-industry-production/farm-classes.

Beef & Lamb New Zealand. (2016c). Whole farm analysis New Zealand. Wellington: Beef & Lamb New Zealand Economic Service. Accessed at www.beeflambnz.com/information/on-farm-data-and-industry-production/sheep-beef-farm-survey/nz.

Bray, A. (2004). More lambs from less. *Primary Industry Management*, 7, 32–33.

Carter, A.H., & Cox, E.H. (1982). Sheep breeds of New Zealand. In G.A. Wickham & M.F. McDonald (Eds), *Sheep production, vol. 1: breeding and reproduction.* Auckland: Ray Richards.

Corner-Thomas, R.A., Kenyon, P.R., Morris, S.T., Greer, A.W., Logan, C.M., Ridler, A.L., Hickson, R.E., & Blair, H.T. (2013). A study examining the New Zealand breed composition, management tool use and research needs of commercial sheep farmers and ram breeders. *Proceedings of the 20th conference of the Association for the Advancement of Animal Breeding and Genetics*, 20, 18–21.

Dwyer, C.M. (2008). *The welfare of sheep.* Dordrecht: Springer.

Ferguson, D.M., Schreurs, N.M., Kenyon, P.R., & Jacob, R.H. (2014). Balancing consumer and societal requirements for sheep meet production: an Australasian perspective. *Meat Science*, 98, 477–83.

Kenyon, P.R., Maloney, S.K., & Blache, D. (2014). Review of sheep body condition in relation to production characteristics. *New Zealand Journal of Agricultural Research*, 57, 38–64.

Mackay, A., Rhodes, A.P., Power, I., & Wedderburn, M.E. (2011). Has the eco-efficiency of sheep and beef farms changed in the last 20 years? *Proceedings of the New Zealand Grassland Association*, 73, 119–24.

Meadows, G. (1997). *Sheep breeds of New Zealand.* Auckland: Reed Publishing.

Ministry for Primary Industries. (2012). Pastoral input trend in New Zealand: a snapshot. Accessed at www.mpi.govt.nz/document-vault/4168.

Morris, S.T. (2013). Sheep and beef cattle production systems. In J.R. Dymond (Ed.), *Ecosystem services in New Zealand: conditions and trends* (pp. 79–84). Lincoln: Manaaki Whenua Press.

Morris, S.T., & Kenyon, P.R. (2014). Intensive sheep and beef production from pasture — a New Zealand perspective of concerns, opportunities and challenges. *Meat Science*, 98, 330–35.

Morrison, I. (1980). *New Zealand sheep and their wool.* Wellington: New Zealand Wool Board.

New Zealand Government. (2010). Animal Welfare (Sheep and Beef Cattle) Code of Welfare 2010.

Rattray, P.V., Brookes, I.M., & Nicol, A.M. (2007). Pasture and supplements for grazing animals. Occasional Publication no. 14. Cambridge, New Zealand: New Zealand Society of Animal Production.

Southby, A. (2008). *Pocket guide to the sheep breeds of New Zealand.* Auckland: New Holland.

West, D.M., Bruere, A.N., & Ridler, A.L. (2009). *The sheep: health, disease and production*, 3rd edition. Vet Learn New Zealand.

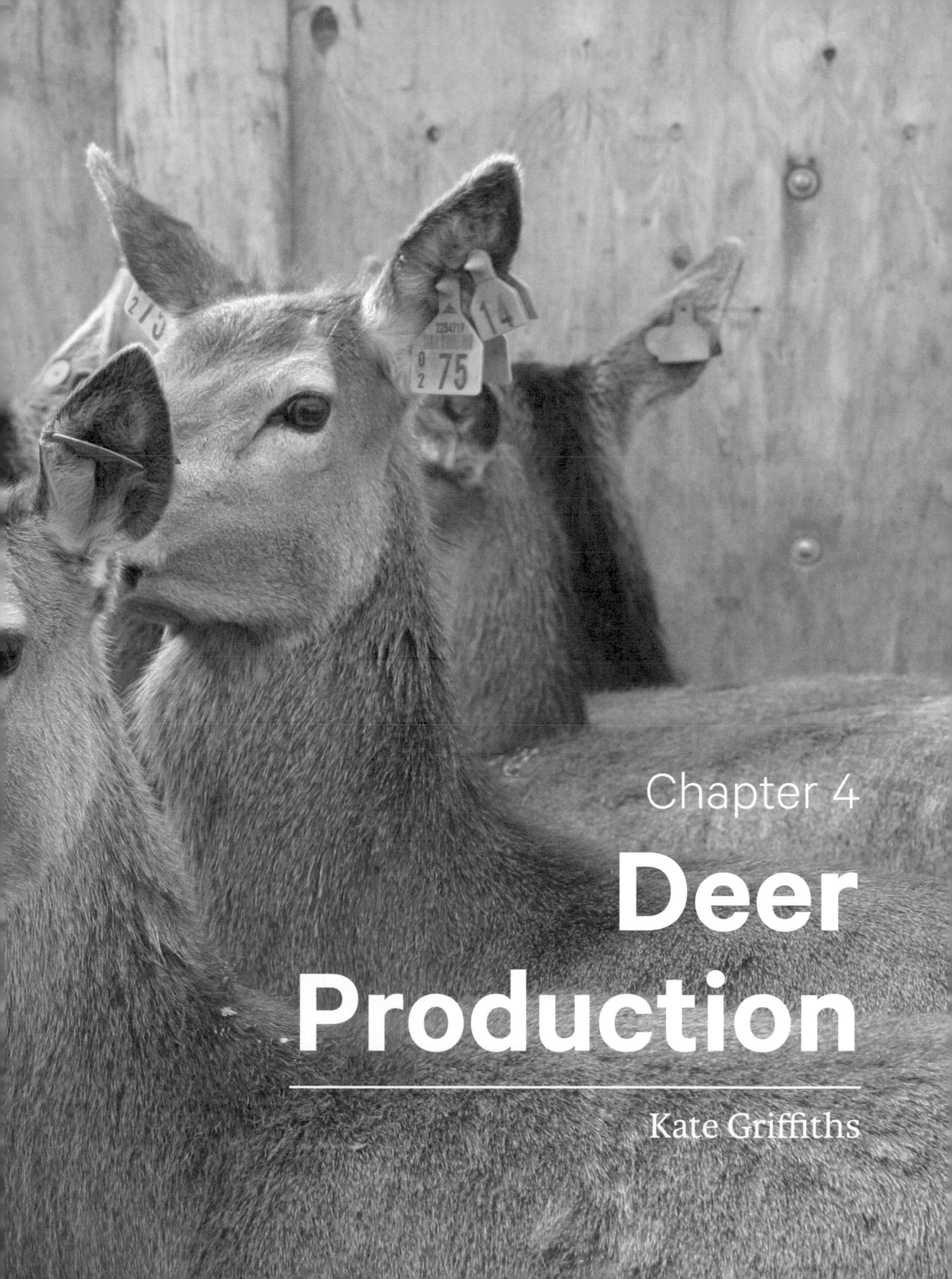

Chapter 4

Deer Production

Kate Griffiths

Chapter 4

Deer Production

Kate Griffiths

Institute of Veterinary, Animal and Biomedical Sciences
Massey University, Palmerston North

Introduction

Seven species of deer live in the wild in New Zealand (Table 4.1), three of which (red, wapiti and fallow) have been farmed successfully. Deer were introduced into the country as game animals in the mid to late nineteenth century, mainly in the Southern Alps and its foothills. The New Zealand environment was ideal for deer and the wild population grew quickly; while some species have remained localised, others have spread widely through one or both main islands. Wild deer are considered to be pest animals due to their impact on the environment,

Table 4.1 **Distribution of deer species in New Zealand**

Common name	Latin name	Distribution
Red	*Cervus elaphus*	Widespread in North and South islands
Fallow	*Dama dama*	Widespread in North and South islands
Wapiti (elk)	*Cervus canadensis nelson*	Fiordland
Sika	*Cervus nippon*	Central North Island
Sambar	*Rusa unicolor*	Manawatu, Bay of Plenty
Rusa	*Cervus timorensis*	South of Rotorua
White-tailed	*Odocoelius virginianus borealis*	Stewart Island, Otago

Source: King (2005) and Department of Conservation.

and are hunted recreationally as game animals. Some species (red and wapiti) are also hunted commercially for the wild venison market.

This pest animal became an export earner in the 1960s, when the export of venison from wild deer began. Early pioneers in the industry saw the earnings and industry potential, and began capturing live deer from the wild with the intention of establishing commercial deer farms. Wild deer were captured in traps on farms that bordered bush areas and by using helicopters that allowed the deer to be netted or darted (Challies, 1985); the majority of the wild deer were caught by helicopter. Freshly caught deer were placed on farms that were fenced to prevent escape, and other facilities to restrain deer were developed. Yards and sheds were designed to allow farmers to handle, draft, weigh and drench deer, and to facilitate velvet harvesting.

Deer became classed as farm animals in 1969, giving rise to the deer farming that we recognise today. Venison and deer velvet became new commercial farm products, being exported mainly to Europe and Korea, respectively.

The deer industry grew until the early part of the twenty-first century, when animal numbers started to decline. The industry's future is unknown; however, the number of farmed deer has remained relatively stable since the mid-2000s, with just under 1 million deer currently being farmed in New Zealand (Table 4.2). Red deer and wapiti, and their crosses, remain the dominant farmed species; the number of farmed fallow deer has declined to a very low figure. There are small numbers of other species on some farms, but they are not commercially significant.

Table 4.2 **Numbers of farmed deer in New Zealand**

Year	Approximate number
1982	150,000
1991	1.2 million
1997	1.7 million
2007	1.4 million
2010	1.1 million (South Island 791,000; North Island 332,000)
2015	0.9 million

Source: Agricultural Production Census.

Wapiti stags. **Previous:** Red deer hinds.

Handling of deer

Designs for deer yards and sheds (Figure 4.1) were developed independently of existing farming systems, as deer are mustered, shifted and handled differently to other farmed species of livestock in New Zealand (Kilgour & Dalton, 1984). Slaughter plants were also developed to cope with these new species (red, wapiti and fallow), each with their own particular needs. Enclosed trucks were used to transport deer. Dark pens with lowered ceilings, which appeared to reduce deer activity, were used to facilitate the safe handling of red deer. Gradually, the design of yards, races and crushes was refined; today, hydraulic crushes are common and high-walled pens and purpose-built sheds allow reasonably safe handling of red and wapiti deer and their crosses. Fallow deer require different shed designs.

The fencing surrounding deer farms is about 2 metres high; internal fencing may be slightly lower. Fencing along raceways needs to be high and strongly built, to withstand the pressure on such fences as deer are moved through. Races at the entrances of yards may be boarded up to encourage forward movement and prevent escape or injury.

Fenced races entering the yards.

Deer yards are boarded up; they often contain a circular central drafting pen from which the deer are moved into smaller rooms or pens. The yards are designed to have no gaps for the deer to get their feet stuck into, smooth edges with no sharp corners, no protruding objects, solid and clean flooring, and solid gates with quick-release latches that open both into and out of the pens.

When handling deer, it is important to consider both the safety of the deer and that of the people handling them. A circular pen with a forcing wall can be used to put pressure on deer, to encourage them to move into a race and thence into a crush. Crushes vary in design, but many are now hydraulically driven with padded walls that can contain (and restrain) several deer at a time.

Deer are routinely yarded for ultrasound scanning of females for pregnancy diagnosis, weaning and tagging of young animals, vaccination, drenching and weighing of stock, and drafting of stock for sale and for removal of velvet from mature males. Tuberculosis (TB) testing of commercially farmed deer is also required, although the frequency of such testing varies between farms. In general, deer are yarded as infrequently as possible to minimise the injuries occasionally incurred during mustering, yarding and restraint.

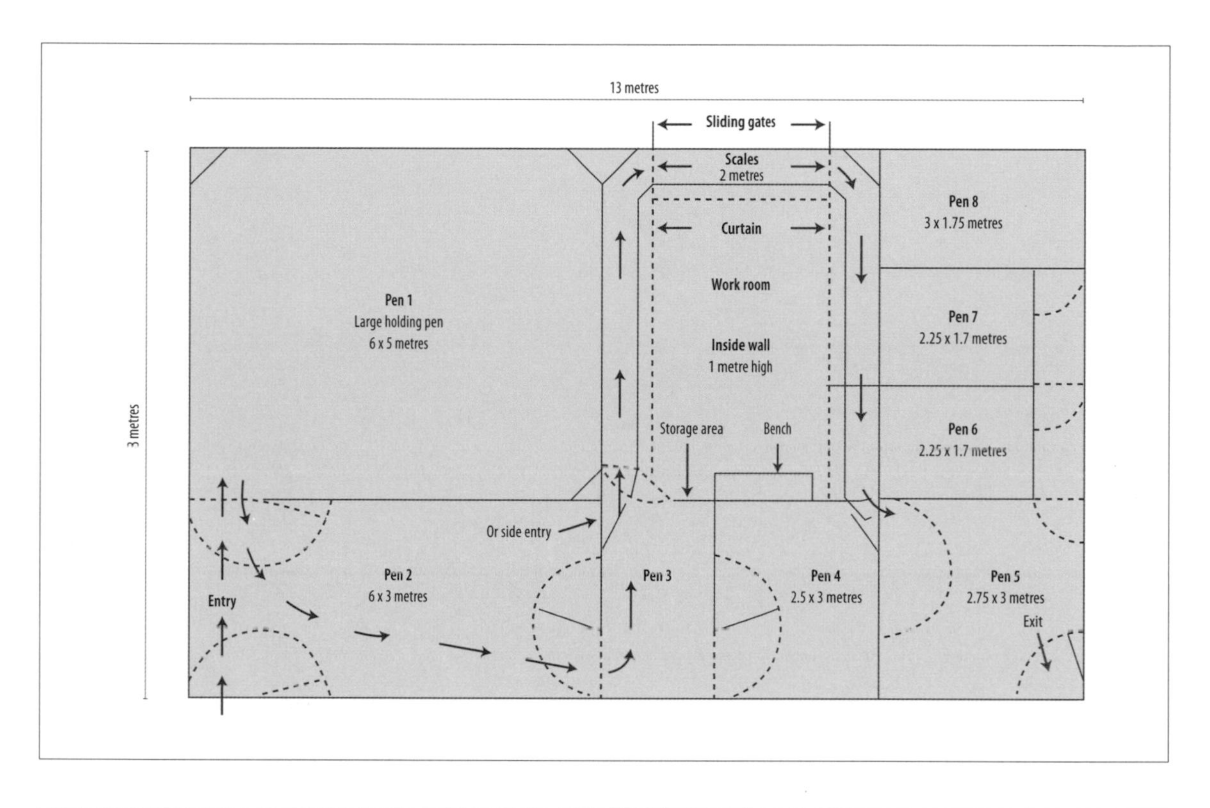

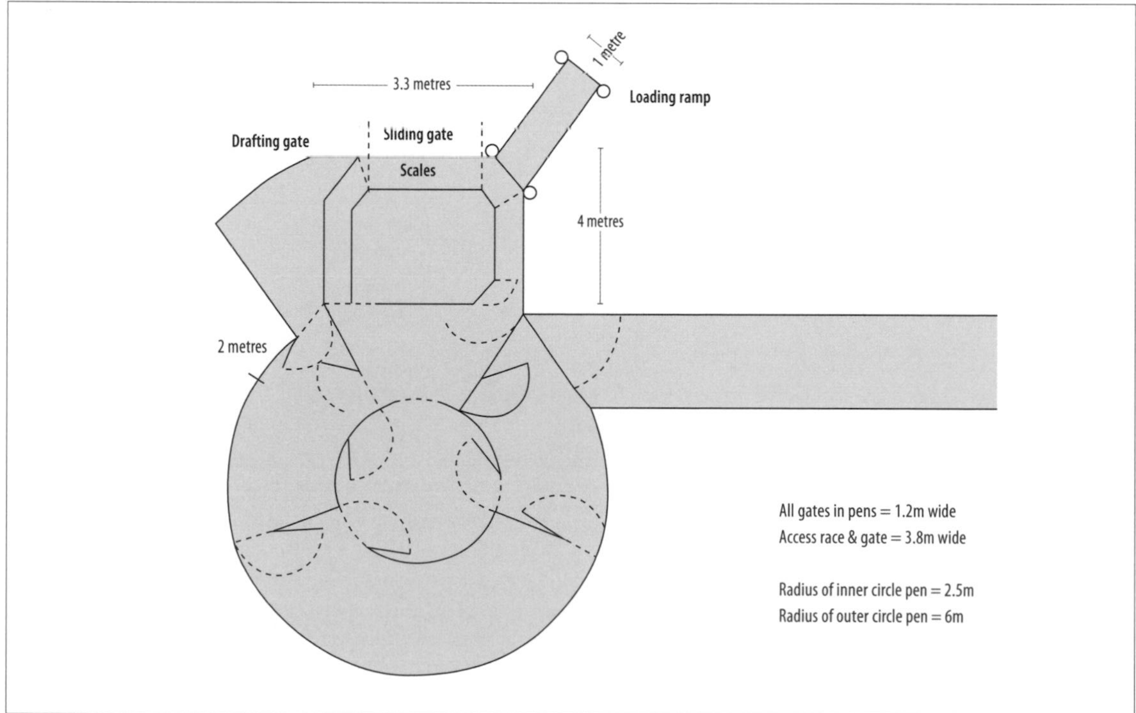

Figure 4.1 Typical designs for deer-handling pens. Source: AgriQuality Ltd (2005).

Glossary

Buck, doe, fawn: terms used for fallow deer (male, female, young).

Bull, cow, calf: terms used for wapiti (elk) (male, female, young).

Roar, bugle, grunt: male deer make noises during the rut (mating season). These noises have different terminology depending on the species of deer: stags roar, bulls bugle and bucks grunt or roar.

Rut: mating behaviour of male deer; also mating season.

Spiker: term used for young male deer.

Stag, hind, calf: terms used for red deer (male, female, young); often used in general for all species.

Velvet: the velvet-like skin covering the soft antler as it grows through the spring and summer.

Red deer stag.

Deer production

The two major products of deer are meat (venison) and velvet. Venison is sold worldwide, although the major market is the German winter market (30 per cent) followed by the Benelux countries (Belgium, the Netherlands and Luxembourg, 23 per cent), the United States (14 per cent) and other European countries (23 per cent) (DINZ, 2016a). Venison from yearling animals is preferred; some meat from older animals is also marketed. Venison is a low-fat highly lean meat and requires cooking techniques that are different to those for prime beef or lamb. It remains popular in traditional markets and has also seen some expansion into recently developed markets. However, the volume traded has declined from 33,000 tonnes (hot carcass weights) in 2006–07 to less than 22,000 tonnes in 2014–15 (DINZ, 2016a). Deer farming for venison does take place in other countries, but on a smaller scale than in New Zealand.

Velvet is sold mainly into Korea and China, where it is used in traditional medicines and tonics. Velvet is taken from stags and bulls, both breeding animals and those farmed specifically for velvet. In other countries (e.g. Russia, China), velvet is harvested from both farmed and killed deer; however, New Zealand velvet is highly valued due to its consistency and quality. Velvet has traditionally had a volatile market, with large fluctuations in the prices paid per kilogram of velvet produced (Figure 4.2).

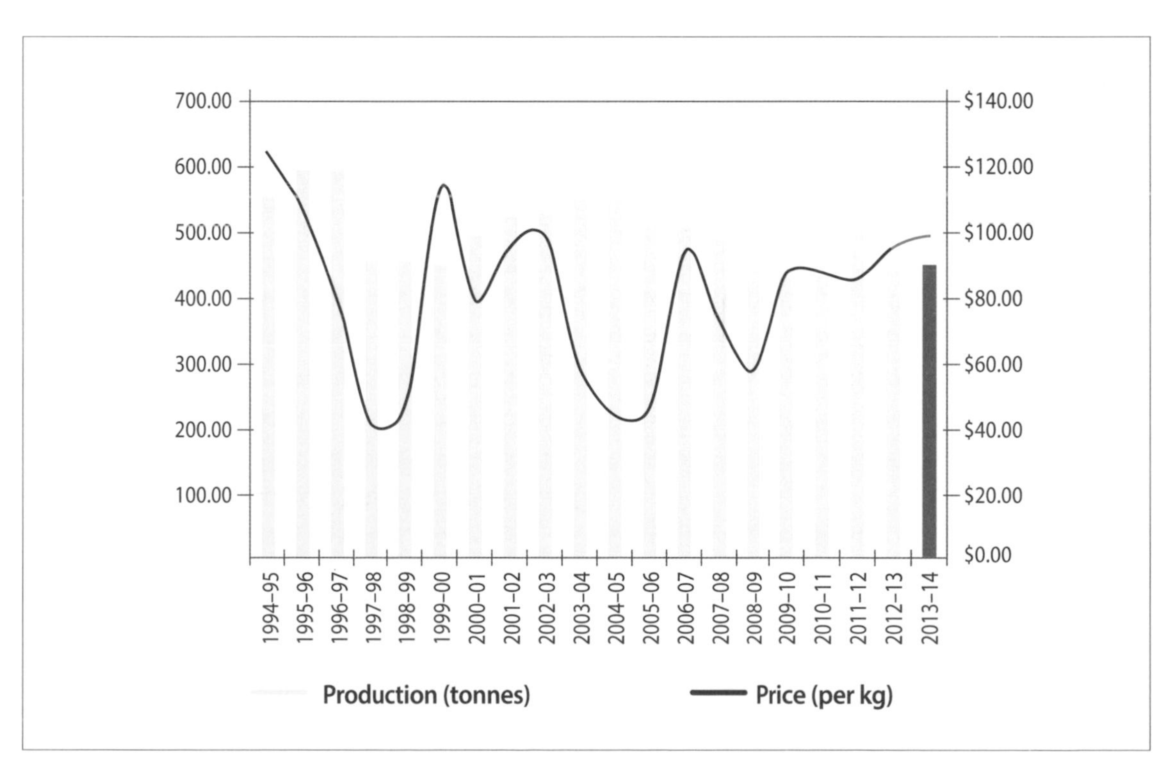

Figure 4.2 Production levels and prices for New Zealand velvet, 1994/95–2013/14 (values for 2013/14 are estimated. Source: DINZ (2016a).

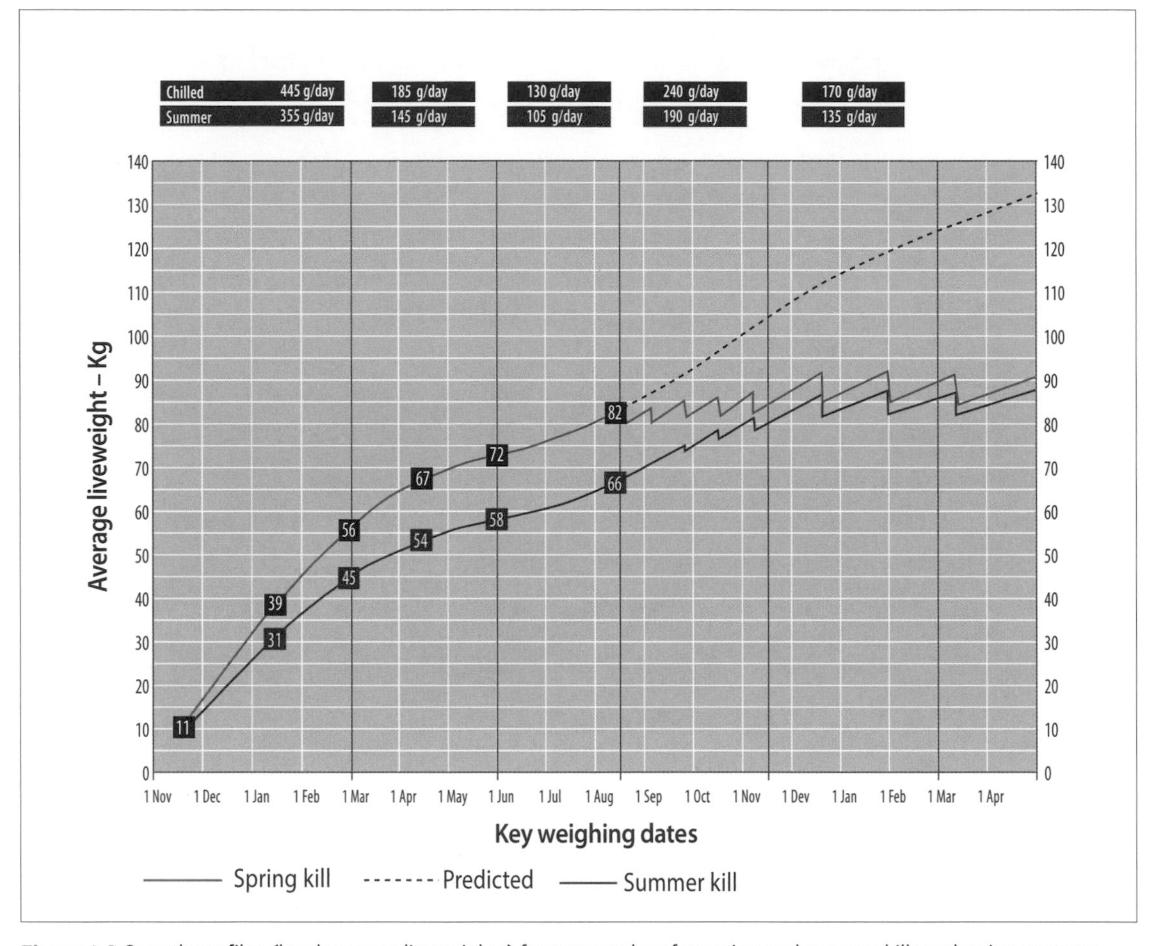

Figure 4.3 Growth profiles (herd average liveweights) for weaner deer for spring and summer kill production systems, based on a 95 kg cut-off point. Source: DINZ (n.d.b).

Deer are farmed on a variety of land types. In general, deer are bred on hard hill country and finished on easier country, but deer farms of different types are located on all classes of land. Sheep and beef farms often have a deer-fenced area.

Deer farms may be categorised into six different types according to the species farmed and the product produced:

- **Red deer and wapiti breeding farms:** these produce weaners, both as replacement animals and for sale. They may be extensive properties on hard hill country, and stock may be yarded infrequently.
- **Red deer and wapiti breeding and finishing farms:** these produce weaners and finish some or all of them for slaughter. The farms may be on easier hill country, with more intensive management of stock and more regular mustering than type 1.
- **Finishing farms:** these purchase red deer/wapiti weaners and grow them to slaughter. These farms are often on good country that allows the cultivation of feed crops. Some farmers finish weaners indoors over winter, but this is very uncommon.
- **Velvet-producing farms:** these farm stags and bulls specifically for velvet production,

Figure 4.4 Seasonal variation in venison pricing and production. Source: DINZ (n.d.a).

and are often situated on easier country.

- **Fallow deer farms:** these breed and finish fallow deer; there are few fallow deer farms in New Zealand.
- **Trophy-hunting farms:** these purchase (and may also breed) mature stags and bulls and allow trophy hunting. Large fenced hunting blocks on rough country give the hunters the impression of hunting on wild country.

The target of deer farmers with breeding hinds is to have more than 95 per cent of adult hinds and more than 90 per cent of two-year-old hinds in calf, and for their calves to survive and grow well to weaning. Well-grown weaners will continue to grow after weaning (typically in March/April) before growth rates drop off in winter (Figure 4.3), then grow well again in spring so as to be large enough for slaughter at just under 12 months of age. Calves may be crossbred from red deer hinds or red x wapiti hinds mated to wapiti stags; the latter are used as terminal sires with most of their progeny going to slaughter.

Deer calves are born in November, and slaughter prices (venison schedules) are at a peak during October to November (Figure 4.4). To maximise returns, deer farmers will thus grow young deer, weaned at four or five months of age in March/April, to slaughter weight at about 11–12 months of age.

Velvet is harvested from stags used as breeding animals, from growing stags and from stags farmed specifically for velvet. Velvet is harvested each year between September and January. Old stags that have been farmed for velvet may then be sold on to trophy-hunting properties for commercial hunting.

Calendar of deer farming operations

The typical calendar of events on deer farms follows the biological changes in male and female deer. Males go through a process of losing (casting) antlers, growing velvet, cleaning (stripping) antlers, holding hard antlers, the rut (breeding season) and again losing antlers (Figure 4.5). Females are mated, become pregnant, calve, lactate, wean their offspring and are then mated again.

Red deer hinds typically calve in November. Most hinds will become restless a few days prior to calving and may show other changes in behaviour. Calving itself is generally a rapid event; dystocia (calving difficulty) is rare in hinds compared with cattle. Calves will usually seek the udder within 30 minutes of birth, after which they hide (seek shelter and camouflage in long grass or brush). It is important that paddocks have calf-proof fences; otherwise, calves may leave the paddock to hide and can become permanently separated from their dams. The hind will visit her calf and feed it two to three times a day. After two or three weeks, the calf will follow its mother and join the herd.

On finishing farms, rising one-year-old deer will be at slaughter weight and will be killed this

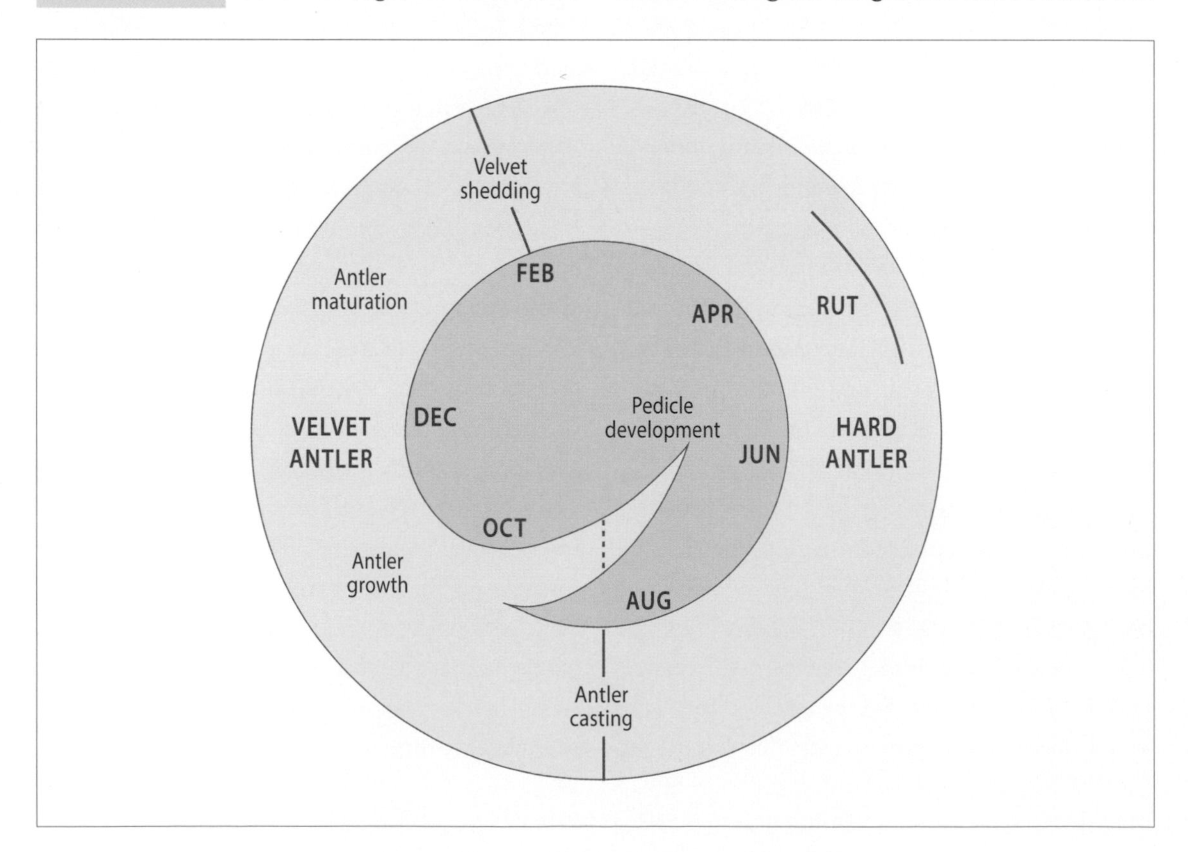

Figure 4.5 Annual antler growth cycle for red deer stags. Source: Fennessy & Suttie (1985).

month. The venison schedule is usually still reasonably good.

Velvet is harvested from stags. Velvet yield is influenced by genetics, and stags are selected for velvet production based on breeding and yield as two-year-olds. Velvet harvesting is defined as a controlled surgical procedure and is carried out under a strict protocol with mandatory pain relief. It must be performed in accordance with the Code of Recommendations and Minimum Standards for the Welfare of Deer During the Removal of Antlers (National Animal Welfare Advisory Committee, 1992).

Velvet can only be removed by veterinarians or approved farmers operating under veterinary supervision. New Zealand has a National Velveting Standards Body (NVSB) that implements (and audits) the standards. Stags must be restrained for velvet antler removal. This can be either physical restraint in a crush (often hydraulic) or chemical restraint (sedation). A tourniquet is placed around the base of the antlers and local anaesthetic is administered in a ring block. The efficacy of the anaesthetic is checked, after which the velvet antlers are sawn off. NaturO rings (specialised rubber rings that provide analgesia through compression) may be used to harvest velvet from one-year-old (spiker) stags only.

The velvet is harvested when it reaches an optimal stage of growth, at approximately 55–65 days after stags last lost their antlers (commonly referred to as button cast/drop). Button casting usually happens in August to September. A farmer may harvest velvet from only a few stags every couple of days (i.e. not all stags have their velvet antlers harvested at the same time), to ensure that it is at the optimal stage (thus maximising its value). Velvet is categorised into a range of grades according to length and width. After removal, velvet is individually tagged and frozen and is then sold to specialist traders. The velvet is then dried and either exported as raw velvet or processed before export. The price for velvet differs between years, from as low as $40/kg in the late 1990s, $50/kg in 2008, $100/kg in 2011 and approximately $120/kg in 2015. The volume exported fluctuates between 400 and 600 tonnes per annum and is increasing.

DEC

Velvet harvest continues. Calving may continue into early December. Hinds are lactating.

Velvet left on stags starts to harden. Hinds are lactating. Stags may be put in with two-year-old hinds this month to allow them to become socialised before mating starts in February to March.

Weaning takes place; weaners are either sold (directly off farm or at weaner sales) or held on farm for growing out as replacement stock or for slaughter. Weaning generally occurs before stags are put out with the hinds for breeding (pre-rut weaning), although a small number of farms may wean after breeding (post-rut weaning). Management of young weaners immediately before and after weaning is critical to ensure that stress is minimised. Animal health treatments such as vaccinations and anthelmintics (drenches) are often administered just before or at weaning.

Mesopotamian fallow hind.

Breeding begins. The stag to hind ratio is usually 1:30 to 1:50 for mature hinds and 1:10 for R2 (rising two-year-old) hinds. Farmers may mate more hinds to particular stags, particularly those with very good genetics, although this risks some hinds not becoming pregnant if that stag has reduced fertility. Hinds can be single-sire mated (one stag with a small group of hinds) or multiple-sire mated (multiple stags with a large group of hinds). The hinds are usually left with the stags for two to three cycles (one cycle is 18–21 days).

With single-sire mating, a second (back-up) stag can be put in with the hinds during the second cycle to reduce the risk of non-pregnancy if the first stag is infertile. Multiple-sire mating is commonplace in New Zealand, although it is important that the stags have been run together prior to mating. Farmers must also be aware of which stags are dominant, and ensure sufficient room in each paddock for each stag to establish a harem of hinds.

Mating mobs should not be kept in adjacent paddocks, as the stags will spend time fence-pacing and being aggressive rather than mating. Two-year-old stags can be used for mating; these may be less aggressive towards each other. It must be remembered that stag behaviour changes over the mating period. As testosterone levels increase, they can become more aggressive towards people, other stags and other animals.

APR-MAY

Mating ends, and stags are removed from hinds. Ongoing animal health monitoring and treatment/supplementation is carried out as required.

Young animals (weaners) should be growing well through autumn, and may be grown on their farm of origin or sold at weaner sales.

JUN-SEPT

Hinds are scanned for pregnancy and the age of fetuses may be determined. Dry (non-pregnant) hinds are culled. Pregnancy scanning allows the calculation of mating performance and an indication of mating and expected calving patterns.

In winter, voluntary food intake is reduced; the deer tend to hold or lose body weight but not grow.

Antlers are cast in August; velvet starts to grow. Poor nutrition in autumn/winter and spring can reduce velvet production. Extra feed in spring can increase velvet growth substantially.

SEPT-OCT

In spring, increased grass growth allows weaners to be fed well to maximise growth for finishing as yearlings. Optimal carcass weights are approximately 50–60 kg (92–118 kg liveweight) for the German winter market (dressing out of approximately 54 per cent). To reach these target weights it is important to ensure that calves are born early, in a compact calving period, and are grown to a good size prior to weaning with growth rates of greater than 400 g/day being achieved. In early March, female weaners should be at a target liveweight of 56 kg with males at 60 kg (red deer). Post-weaning feeding needs to be well managed to maximise weight gains in autumn and spring when the animals have maximum growth potential.

As voluntary food intake decreases during winter, growth then slows in weaners. From birth to weaning, calves might gain 400+ g/day; from March to May, weaners typically gain 100–180 g/day; weight gain then drops to 50–150 g/day from May to September and increases again to 220–250 g/day in September to December.

Velvet harvesting begins.

Rising two-year-old wapiti/red cross stags.

Nutrition

A potential problem with the commercial farming and breeding of deer in New Zealand is a mismatch between feed supply and feed demand (Figure 4.6). Grass growth in New Zealand follows a typical pattern of increased growth in spring (spring flush), reasonable growth in summer and autumn (although this is area-dependent) and reduced growth over winter. Deer-only breeding systems are often unable to fully utilise the spring flush, as the hinds' demand for feed has not yet increased sufficiently. It is important that this pasture is utilised (e.g. by other classes or species of stock or through making supplements) so that pasture quality is maintained going into summer when hinds are lactating.

In some areas, reduced feed supply and quality can become an issue in late summer; this is of particular concern as there are weaners to be grown and hinds are approaching the breeding season. It is becoming increasingly common to utilise alternative forages such as chicory and clover mixes to feed young growing deer during this time.

Deer are maintained on pasture; this is usually ryegrass and clover mix swards, but on hard hill country the sward may be poorer with brown top and brush included. The optimum length of pasture for deer is approximately 10 cm, with a residual height of 7–8 cm optimal and pasture quality remaining high (greater than 11 megajoules of metabolisable energy per kilogram [MJ ME/kg]). Grazing management must be considered, with pasture covers and

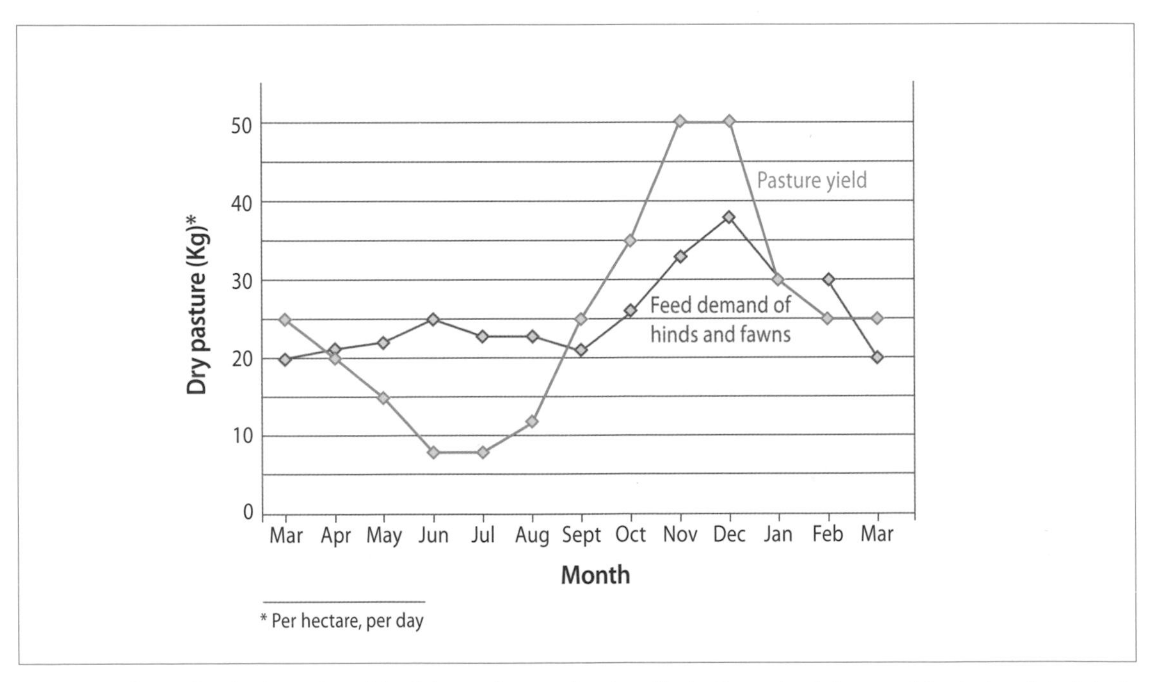

Figure 4.6 Illustrative feed supply and feed demand curves for deer, illustrating potential areas of feed mismatch. Actual curves will vary by farm location and type. Source: DINZ (2016b).

Wapiti stag with antlers — velvet drying up.

rotation lengths monitored. Integrated grazing management (i.e. with other species) can work well on deer farms, and can help maintain pasture quality due to the different demands and grazing patterns of different species.

On better country, deer may be fed on crops to increase growth, improve performance or to ensure that feed shortages do not occur. Crops grown for deer include swedes, fodder beet, kale, rape, chicory, plantain, clover and lucerne; what is grown depends on the weather, land type and type of stock.

Stags that are kept for velvet production may be fed grain or high-quality supplementary feeds in late winter to spring to improve performance. Deer must be transitioned gradually onto supplementary grain. On some farms, hinds and their calves may be fed supplementary grain for a couple of weeks prior to weaning, with the calves continuing to be supplemented following weaning. This can help to reduce the stress often associated with weaning.

Breeding and genetics

To improve the quality of farmed deer, genetics from wapiti from North America, red deer from Eastern Europe and fallow deer from England, Hungary, Sweden, Germany and Denmark have been introduced.

In New Zealand, farmed deer become sexually active in the early autumn. Red deer hinds reach puberty at about 16 months of age and come into oestrus every 18 days for about 24 hours. Red deer stags reach puberty at about 12 months. In the wild, stags and hinds live separately until the rut when stags join hinds and mating occurs.

Wapiti stag: note the height of the fence.

Stags roar to challenge other stags. During the rut, stags' antlers are used for demonstrating size and ability and for fighting with other stags.

Some calves bred for venison are from hinds that are a cross of wapiti and red deer, and the stags are often wapiti. This is to maximise hybrid vigour and produce larger calves, which typically have increased growth rates enabling target slaughter weights to be attained earlier. The gestation period for red deer hinds is 229–232 days.

Artificial insemination and embryo transfer are used in deer, with high pregnancy rates achievable. If hinds are to be artificially inseminated they must be synchronised, as there is currently no reliable method of heat detection available for use in farmed deer.

A number of stud breeders exist in New Zealand; they keep records of individual animal performance and select animals to breed from based on a range of desired traits. This information can be provided to Deer Select, New Zealand's national deer recording database. Deer Select stores pedigree and performance (trait) records, and then uses the data to provide estimated breeding values and economic indices. Deer stud breeders can use these breeding values and indices to make selection decisions and monitor their genetic progress within the nationally recorded herd (i.e. how their animals compare with others). Commercial farmers can use breeding values and indices to select stud breeders and make purchasing decisions based on directly comparable data across different stud herds. The velvet and venison selection criteria are separate; each selected individually so as to return the best gains.

Animal welfare

The major welfare issues of deer are velvet harvesting, transport, weaning, overreacting and being injured during yarding and restraint (Wilson & Stafford, 2002). Many deer are still afraid of humans and overreact when in close proximity to them, such as when they are yarded and exposed to various management procedures. Injuries to deer can be significant; one study in the early 2000s found that up to 15 per cent of weaner mortalities were due to injuries. Deer can be desensitised to the presence of humans, such as during supplementary feeding with maize. It is likely that deer are inadvertently being selected for tolerance of humans through the loss of those that overreact and by the selective culling of any that are particularly aggressive towards humans.

Deer may pace along fences, particularly after capture, during the rut and before calving, and this may be increased by grazing in small paddocks with no cover. This behaviour was particularly obvious in the early years of deer farming but has reduced over the past few decades. The improvement may be due to changes in farm practices and a change in deer responses to being confined to paddocks. It may be a sign of the increased domestication of commercially farmed deer.

Wallowing red deer stag.

Weaning is stressful, particularly if done early before the rut. Minimising this stress in weaners is important to reduce the risk of injury and/or disease. Stress may be reduced by strategic feeding of grain, placing weaners back into the paddock they occupied with their dams before weaning, avoiding mixing weaners from different fawning paddocks, and the use of 'Nanny' hinds. It is essential to provide recently weaned calves with high-quality feed.

Velvet harvesting is a major welfare issue for deer. Velvet is sensitive tissue that regrows annually and its removal is certainly painful. Local anaesthetic must be used to control the acute pain experienced during removal of the velvet. There is little information on the long-term pain caused by velvet harvesting, or its control.

The welfare of farmed deer is covered by several codes of welfare, including the Animal Welfare (Deer) Code of Welfare 2007, the Code of Recommendations and Minimum Standards for the Welfare of Deer During the Removal of Antlers, and the Code of Welfare for Transport of Animals in New Zealand. These codes are available on the MPI website (www.mpi.govt.nz/protection-and-response/animal-welfare/codes-of-welfare).

Disease

The major health problems of deer include TB, yersiniosis, fusobacteriosis, leptospirosis, Johne's disease, malignant catarrhal fever (MCF), copper deficiency, internal parasites and ticks (Wilson, 2002). Facial eczema and ryegrass staggers are also a problem in some areas. Dystocia is a minor problem in hinds (typically less than 1 per cent). Post-capture myopathy (disease of muscle tissue) was a problem when deer were initially captured and farmed; it is now less commonly seen. Injuries due to misadventure are common in deer, especially during mustering and yarding, but the number of such injuries has decreased as deer have been selected for greater quietness and less aggression and handling facilities have been improved. The incidence of TB is declining due to a widespread compulsory testing scheme.

Yersiniosis is a disease caused by the bacterium *Yersinia pseudotuberculosis.* It is common in young (weaner) deer during their first autumn/winter and presents as acute (sudden) diarrhoea or sudden death. Surviving deer may experience a period of reduced growth. The biggest risk factor for yersiniosis is stress, such as that following weaning, transport, inclement weather or poor nutrition. The risk of yersiniosis can therefore be reduced by appropriate management of potential stressors, particularly around the time of weaning. A yersiniosis vaccine is also available; this should be given prior to an expected period of stress and requires a sensitiser and a booster dose before it is effective.

Fusobacteriosis is a bacterial disease that causes purulent, necrotic lesions on the feet and/or mouth that may then spread to the lungs or liver. The bacterium gains entry through damaged skin, for example caused by abrasions or trauma or cold, muddy conditions. To reduce the risk of infection, facilities such as transport trucks and sheds/deer yards need to be well maintained, well designed and routinely cleaned. Careful deer handling is important to avoid injuries and overcrowding.

Leptospirosis (lepto) is an important bacterial disease of farmed deer. There are various strains of lepto that can cause subclinical disease (e.g. reduced growth or reproductive performance) or clinical disease (sick and dying deer). There are also zoonotic risks, as leptospires shed in the urine of infected animals can infect humans. Vaccines are available for use in deer. It is also important to maintain good hygiene when handling deer, such as wearing gloves, covering any cuts and washing hands.

Johne's disease causes scouring, weight loss and death in deer. It is usually seen as sporadic cases in mixed-age deer, although outbreaks of disease can occur in young deer. Johne's disease has caused significant problems in the industry, resulting in the formation of Johne's Management Ltd (JML) which is involved in ongoing surveillance and monitoring and prevention programmes on farms.

Malignant catarrhal fever (MCF) is a viral disease of farmed deer in New Zealand; it is caused by ovine herpesvirus-2, which is associated with sheep. It is a sporadic disease causing eye, oral and gut lesions and is fatal.

Copper deficiency can be seen in farmed deer, presenting with a range of possible signs including ill-thrift (slower-than-expected growth or

Red deer hinds in yards.

Young red deer stags with velvet.

loss of condition), bone fragility, osteochondrosis (degeneration in the joints of the long bones) and swayback (lack of coordination and swaying on the hind legs). It is essential that the copper status of deer on each individual farm is assessed, enabling the implementation of an appropriate copper supplementation plan. Options for copper supplementation include fertiliser, oral drenching (not practical in deer due to the frequency required), intra-ruminal boluses (copper bullets) and copper injections. Copper is toxic in overdose, so care with supplementation is required.

Internal parasites are an important health issue in farmed deer, with increased incidence on intensive and finishing farms. Lungworm is very common, particularly in young deer in the autumn/early winter; it can present as spectacular outbreaks of sudden death with little prior warning. Gastrointestinal (GI) parasites are also important in deer. There are a range of different GI parasites, and if uncontrolled these can cause reduced growth through to clinical disease and, in severe cases, death. A number of different options are available to aid in parasite management, including cross-grazing with other species, alternative forage use and anthelmintic use.

In certain areas in the North Island, ticks can cause the death of newborn calves, damage hides and velvet antlers, and reduce growth/production. They can be controlled via pasture management or chemical treatments.

Conclusion

The deer industry has contracted over the past decade as farmers have exited the industry. Good deer-finishing country has been converted into dairy farms or used for dairy support. Improved beef, lamb and dairy prices and poor venison prices will, naturally, influence stock classes on a farm. Moreover, the change in farmer generation may have had an effect, with those who were involved in the establishment of the industry retiring and being replaced by farmers with less emotional investment in the farming of deer. However, those who are in the industry are positive about its outlook, with the future of deer farming dependent on adequate prices for the high-quality product New Zealand farmers produce.

References

AgriQuality Ltd (now AsureQuality). (2005). *Safe practical deer yards.* Wellington: AgriQuality Ltd.

Challies, C.N. (1985). Establishment, control and commercial exploitation of wild deer in New Zealand. In P.F. Fennessy & K.R. Drew (Eds), *Biology of deer production* (pp. 23–26). Wellington: Royal Society of New Zealand, Bulletin 22.

Deer Industry New Zealand (DINZ). (n.d.a). DINZ activity/P2P — a deer industry initiative/What is Passion2Profit? Accessed at www.deernz.org/dinz-activity/p2p-deer-industry-initiative/what-passion2profit#.V8Y_dGUbeOo.

Deer Industry New Zealand (DINZ). (n.d.b). Venison production: set your growth targets (wall planner). Accessed at www.deernz.org/sites/dinz/files/Poster%20 P2P-newpic.pdf.

Deer Industry New Zealand (DINZ). (2016a). About the deer industry/Deer industry statistics/Velvet production and pricing trends. Accessed at www.deernz.org/about-deer-industry/nz-deer-industry/deer-industry-statistics/velvet-production-and-pricing-trends#.V7-v4mUbeOo.

Deer Industry New Zealand (DINZ). (2016b). Deer hub/Feeding/Feeding deer/Balancing supply and demand. Accessed at www.deernz.org/deerhub/feeding/feeding-deer/balancing-supply-demand#.V7-yvGUbeOp.

Fennessy, P.F., & Suttie, J.M. (1985). Antler growth: nutritional and endocrine factors. In P.F. Fennessy & K.R. Drew (Eds), *Biology of deer production* (pp. 239–50). Wellington: Royal Society of New Zealand, Bulletin 22.

Kilgour, R., & Dalton, C. (1984). *Livestock behaviour: a practical guide.* Auckland: Methuen.

King, C.M. (2005). *The handbook of New Zealand mammals.* Victoria, Australia: Oxford University Press.

National Animal Welfare Advisory Committee. (1992). Code of recommendations and minimum standards for the welfare of deer during the removal of antlers. Code of Animal Welfare no. 5. Wellington: Ministry of Agriculture and Fisheries.

Wilson, P.R. (2002). Advances in health and welfare of farmed deer in New Zealand. 2. Velvet antler removal. *New Zealand Veterinary Journal*, 50, 105–19.

Wilson, P.R., & Stafford, K.J. (2002). Welfare of farmed deer in New Zealand. *New Zealand Veterinary Journal*, 50, 221–27.

Chapter 5

Goat Production

Colin Prosser and Kevin Stafford

Chapter 5

Goat Production

Colin Prosser
Dairy Goat Co-operative, Hamilton
Kevin Stafford
Institute of Veterinary, Animal and Biomedical Sciences
Massey University, Palmerston North

Introduction

In New Zealand, goats are farmed to produce milk, fibre and meat and are used for weed control. There is a large population of feral goats in the bush, and control of these by professional and recreational hunters is common. The dairy goat industry is developing into a significant farming enterprise, and farming Angora goats for mohair is a small but important industry.

On many stations, goats live in a semi-wild state — usually ignored until they are mustered for slaughter as meat goats. More deliberate farming of goats for meat production is a small niche business. Individual goats — pet goats — tethered by the roadside used to be a common sight along country roads but are now less common and the practice may be prohibited in future.

At present, the mohair industry is fairly static but the dairy goat industry is expanding. Meat from farmed goats is a minor industry, based to some extent on specific meat goat flocks but also using goats mustered off extensive sheep and beef farms.

Dairy goats

Dairy goat farming in New Zealand is a niche farming enterprise, with an estimated 50,000 milking goats. However, exact figures are not known as there is no census of either dairy goat farms or milking goats in New Zealand. Most commercial dairy goat farms are located in the Waikato region and supply the Dairy Goat Co-operative (NZ) Ltd. The size of herds on commercial farms ranges from 210–1800 lactating goats (average 650 goats), but there are many small holdings milking only a few goats. Some of these supply goat milk for on-site processing of boutique cheeses, or fresh goat milk for the local consumer market.

Total global milk production from all species in New Zealand is estimated to be about 739 million tonnes, with goat milk making up 17 million tonnes, or 2 per cent, of this. By international standards New Zealand goat milk production is insignificant, probably representing less than 0.5 per cent of the goat milk produced worldwide. However, New Zealand was the first nation to develop a commercial infant formula made from goat milk, and is still the leading international manufacturer of goat milk nutritional powders for infants and young children. Most goat milk is converted to powder, of which approximately 90 per cent is exported to Australia, South Africa, Asia and Europe (Anon, 2011).

Angora buck. **Previous:** Saanen dairy goats being milked.

A brief history of goat production

Goats were carried on ships in the eighteenth century, and there was one with Cook on his first voyage to New Zealand. He later liberated goats into the Marlborough Sounds; other ships left goats around the coast and on islands as food for castaways. Goats were used widely as stock animals by early settlers, surveyors and prospectors (Parkes, 2004); they later became feral and today are found in various environments on both main islands.

In the 1880s, Angora goats were imported to start a mohair industry. This industry collapsed and the goats were released. In the 1980s something similar happened; during a feral goat eradication attempt in the Kaimai Ranges in the 1980s about 50 per cent of the goats shot had ear tags.

The number of farmed goats has fluctuated over time. In 1984 it was around 200,000, then rose rapidly to over 1.2 million in the late 1980s before collapsing to less than 500,000 in 1994 and probably less than 100,000 in 2008. Sixty-six per cent of these animals were farmed in the North Island, with the largest population in the Waikato and an estimated two-thirds of all goats located on commercial sheep and beef farms (Anon, 2011).

Saanen dairy goats in shed.

The first attempt to establish a milking goat industry in New Zealand was made by the Agricultural Department in 1921. This venture apparently didn't last very long, and the goats were destroyed (Sheppard & O'Donnell, 1979). Smaller herds of goats were maintained by individuals, with goat milk supplied for local consumption.

In the 1970s, commercial production and processing of goat milk into products such as cheese or milk powders was reinvestigated. The limiting factors in establishing a viable industry were the seasonal supply and the poor quality of the milk at that time. Goat milk was often frozen and stored until there was enough milk for processing; the quality of the milk was therefore variable and it generally had a strong goaty flavour. However, the demand in the Pacific and South East Asia for goat milk powder was initially strong enough to drive the growth of several small goat producer groups around New Zealand.

Unfortunately, like many fledgling alternative farming enterprises, oversupply and falling demand for products led to many farmers having to pull out of the industry. By 1984 there were only 22 surviving farms; these went on to form a single co-operative in an attempt to rebuild

Saanen dairy goats entering milking parlour.

the industry and create a market-led company. This co-operative become known as the Dairy Goat Co-operative (DGC). The DGC initially supplied bulk goat milk powders, but its success came from developing the world's first commercial infant formula from goat milk in the late 1980s. From 2003 onwards, the DGC has built spray-drying, dry-blending, can-making and canning facilities on its site in Hamilton.

In 2012, the DGC had the eighth-highest annual turnover of all New Zealand dairy firms, at $118 million. Its annual turnover increased by $30 million from 2011 to 2012, a growth of 34 per cent and the third largest of all New Zealand dairy firms. The five-year compound annual growth rate (CAGR) of annual turnover for the DGC was 16 per cent, highlighting the robust growth of this new market entrant (MBIE & MPI, 2014a). As of 2014, the co-operative had 72 supplying farms. Several other companies based in New Zealand are now also producing goat milk nutritional powders for export into South East Asia.

Milk production

New Zealand operates a seasonal production system. Kidding starts between 10 June and 25 July, and goats are milked for 212–324 days (average 296 days). Dairy goats will generally produce between 660 and 1800 litres of milk in a 305-day lactation period, with a good-quality doe giving at least 2.7 litres/day on average and 3.5 litres/day at peak (Anon, 2011). Dairy goats produce

25–126 kg of milk solids (kg MS) per lactation (average 86 kg); the weight of a mature doe is around 60–90 kg.

Well-fed dairy goats average 200 per cent kidding (two kids per doe), which contributes to the ability for rapid herd improvement and expansion. Well-grown kids are generally mated to start milking at 13 months of age. Goats are capable of lactating for several years without further pregnancies, although most commercial herds are seasonal. Animals may remain in the commercial herd for 5–8 years, or until milk production drops or health problems arise. Male kids have no commercial value at present and are most often culled, representing a significant loss to the system. Only a very limited number of male dairy goat kids are used in meat production, and even then the returns are less than the costs of raising them. Female kids can be reared as replacements or for local and export sale (Anon, 2011).

Composition of goat milk

Goat milk is an excellent source of high-quality, readily digestible proteins and fats, vitamins and minerals. An added advantage of goat milk for infants is that the milk proteins are secreted by a process that is similar to the secretion of human milk. As a result, goat milk contains many of the components that are important for infant growth and development, although it still requires supplementation to produce infant formula. Goat milk has a different composition to cow milk (Ceballos et al., 2009). Goats also have much greater genetic diversity than dairy cattle, with variable levels of alpha-s1-casein in milk (Martin et al., 2002). The low levels of alpha-s1-casein is one of the factors responsible for the soft, crumbly structure of goat cheeses and may explain some of the reports of the different allergenic properties of goat milk. Goat milk does not contain carotene, and hence cheese or butter made from goat milk is very white. Heritability of milk, fat and protein traits are medium to high; thus, herd improvements can be made quite rapidly through breeding schemes (Anon, 2011).

Goat milk in New Zealand is used for fresh or UHT milk, yoghurt, cheeses and nutritional powders for infants and young children. Most of the infant formula is exported overseas, while the other goat milk products are produced only for domestic consumption.

Saanen dairy goats on a rotary milking platform.

Farm systems

There are two dairy goat farm systems in New Zealand. The first is a free-range grazing system similar to dairy cow farms, while the other manages goats off paddock, bringing fresh pasture to the goats two to three times a day. The practice of housing goats in freestall barns (in which they can move around freely) is common internationally. In New Zealand, goats are housed in open-sided barns where they are exposed to natural lighting and fresh air. Adult goats are given a minimum of 3 m^2 per doe to ensure adequate space for feeding, lying down and moving around the barn. Off-paddock farming systems were initially used in New Zealand to minimise the intake of intestinal parasitic larvae by animals grazing directly on pasture.

Goats managed off paddock typically produce 30 per cent more milk than goats kept outdoors — it allows more feed options to offset seasonal pasture production and allows milking all year round. One disadvantage of off-paddock management is that the capital and running costs of the facilities and equipment are high (Solis-Ramirez, et al., 2011); it is also labour-intensive, with labour needed even when the goats are not being milked.

Carbon footprints

The carbon footprints of dairy goat farming in New Zealand have been measured (Robertson et al., 2015). The off-paddock farming system has a lower carbon footprint than the grazed farming system because of the greater milk production per doe. In New Zealand, the dairy goat carbon footprint is of a similar magnitude to the cow dairy footprint per unit of milk yield corrected for fat and protein (Robertson et al., 2015).

Growing dairy goats in shed.

Saanen dairy goats being fed outdoors.

Breeding

The milking goat population in New Zealand is based on European breeds, including Saanen, Toggenburg, Sable, British Alpine and Nubian, with over 80 per cent of milking goats being the Saanen breed (Solis-Ramirez et al., 2011). The New Zealand dairy goat herd is of a high genetic quality and highly regarded by other dairy goat industries around the world (Anon, 2011). Unlike dairy cattle, there is no national genetic improvement scheme and little artificial insemination. The New Zealand Dairy Goat Breeders Association offers a classification and production recording service for all goat owners in New Zealand. The association maintains a register of goats for members, with details of awards, classification scores, production records, transfers and leases. This system is mostly used by owners of small herds. Most commercial farmers use pedigree selection, based on herd testing, to choose replacement does and breeding bucks. Twins are common, and milking goats are often bred at nine months of age. As a result, the potential genetic gains can be rapid. A genetic improvement programme for commercial dairy goat herds has been modelled by researchers at Massey University (Solis-Ramirez et al., 2012), but has not yet been widely adopted.

Feeding

Unlike sheep and cattle, which predominantly graze leafy material, goats are selective browsers and will tolerate browsing of species with large amounts (up to 50 per cent of the dry matter) of tanniferous (tannin-containing) compounds (Silanikove et al., 2010). Grazing goats can quickly turn mixed ryegrass/clover pastures into clover-dominant fields.

In New Zealand, dairy goats are fed forages such as grasses, clover, lucerne, hay, chicory, plantain, silage or other pasture plants that are either grown on farm or bought in from other farms (Table 5.1). While well-managed forages grown on farm remain the cheapest high-quality feed available for dairy goats, there is increasing use of non-forage supplements. It is common for milking goats to be given some meal supplement at milking, although palm kernel extract (PKE) is not permitted as a feed on farms that supply the DGC.

Table 5.1 **Recommended diet composition for milking goats**

Component	% DM consumed
Proportion of forages (fresh and conserved) in the diet	> 50
Proportion of concentrates in the diet	< 40
Proportion of NDF in the diet	> 35
Proportion of forage-sourced NDF in the diet	> 25
Proportion of starch in the diet	< 25

Source: based on recommendations by Legarto & Leclerc (2011) of the French Livestock Institute.
% DM = percentage of dry matter; NDF = neutral detergent fibre.

Mohair and cashmere production

Mohair

Mohair is produced from the Angora goat, which originated in Turkey. Commercial mohair production was established in the nineteenth century in South Africa and the United States. The Angora goats introduced into New Zealand by the government in 1921 came from these countries; it was another 50 years before mohair was exported when, in 1977, the Lands and Survey Department sent a bale of mohair to the United Kingdom (Sheppard & O'Donnell, 1979).

The 1980s were a boom time for the mohair industry. Market prices rose from $8/kg to $18/kg and there were generous tax incentives for producers. As a result, an Angora buck could sell for tens of thousands of dollars, with the top price reaching over $100,000. Due to the tax incentives, a goat producer could buy a goat for $60,000 and write it down in the books for as little as $6. Corporate investors inevitably became involved, causing the number of mohair producers to surge from 50 to 3000 in the 1980s. The mohair industry changed in the early 1990s when New Zealand allowed goats to be imported from Australia. This greatly increased the number of goats in the country and the government

Angora goat.

adjusted the tax laws accordingly, causing the value of goats to plummet.

In 2007 there were about 20,000 Angora goats in New Zealand; at this time, Angora goat farming was more profitable than lamb production. In 2009 New Zealand produced 53,000 kg of mohair, a mere 1 per cent of total world production. The price for adult coarse fibre was $11/kg, while the highest-quality superfine kid fibre attracted $40/kg for 25 micron fibre (Anon, 2011). By 2011, around 800 mohair farmers were producing approximately 80,000 kg of mohair annually.

At present the market for mohair is undersupplied. At its peak, in the 1990s, worldwide mohair production reached 24 million kg per annum; at present, it is just under 3 million kg. Accordingly, mohair prices are greater than those for wool. Prices per kilogram for top-grade kid, kid, young goat and adult fleeces are $32, $28, $21.50 and $17.50, respectively (Mohair Producers NZ, 2016a). A two-tooth doe can sell for $100–$140, with mixed-age does making $115–$130, top bucks $1000 and average bucks $400–$500.

New Zealand mohair growers send fleeces to

Ohuka Farms at Drury in the North Island and/or to Mohair Pacific at Rangiora in the South Island. There, fleeces are classed and sent to South Africa, the world's mohair production hub, where they are processed and exported to the main markets in Italy, Japan and China. Mohair is used in high-end yarns for knits, weaving, embroidery and high fashion or may be blended with Merino wool for high-quality men's suit fabric.

Mohair is a luxury fibre noted for its high lustre that enhances dyed colours; it is smooth, soft and silky to handle relative to its fineness, and warm relative to its weight. Mohair can be bent and twisted without damage to the fibre and is very elastic. As it readily returns to its original size and shape, mohair garments resist wrinkling during wear. Mohair is also dirt-resistant, as dust does not stick to the fibres; this makes it suitable for furnishing fabrics and hangings.

Mohair grows continuously in ringlets of 15 cm or longer with a fibre diameter of 19–45 microns. Its growth rate and diameter vary with the seasons and nutrition, and there is a positive correlation between fleece weight and fibre diameter. Fibre growth rate increases with the age of the animal to peak at 3–4 years of age, and its diameter increases by 8–11 microns from six months of age to maturity.

Mohair is graded according to fineness, length and type. The uniformity of the mohair has a bearing upon its value. There are five lines for mohair: superfine kid fleece (SFK) at 26 microns; fine kid fleece (FK) at 30 microns; young goat mohair (YG) at 32 microns; adult hair (H) at 34 microns; and strong adult hair (SH) at 36 microns. The ideal length for mohair is between 90 and 150 mm. Generally, the longest and coarsest mohair grows on the neck; that on the breech (lower hind legs) may be shorter and contain more kemp. Mohair can contain non-medullated fibres (those without any air-filled internal cells), non-medullated fibres with medullated tips, medullated fibres and kemp fibres (highly medullated, brittle fibres). High-quality mohair contains little or no medullated or kemp fibres.

Angora goats are generally smaller than feral goats, frequently produce twins, and both sexes are horned. Good Angora goats produce 5–6 kg of mohair per year in two shearings, with older animals producing more than young stock (but with fibres having a greater diameter). As shearing time approaches, efforts should be made to protect the mohair from contamination with vegetable matter and dirt as this will devalue the fleece. The goats should be held in well-grassed, open paddocks and not overcrowded in yards. The fineness of the fibre is largely determined by the age of the animals, and drafting the herd into age groups prior to shearing makes classing easier.

An established Angora goat flock is one that is age-balanced — made up of all ages from kids, yearlings (goatlings), two-tooths, four-tooths, six-tooths and five-year-olds (full mouth) and older. Breeders will select for a number of different traits, including fleece weight, fibre diameter, white fibre, lustre and fleece free from hair and kemp. Other desirable traits include early breeding, fast-growing young stock and resistance to worms and footrot. The more traits selected for, however, the slower the progress in any individual trait. The heritability of a trait must also be considered, as this indicates how much of the trait is likely to be passed on to the next generation.

Angora goats are usually farmed like sheep on pastures of ryegrass and white clover. They may be fed supplements in winter or during droughts, and can also feed on browse (shrubs, bushes and trees) which may be added to their diet to supplement grazing. In some situations they may be used to control weeds to improve grazing for sheep and cattle.

Cashmere

Cashmere is a fine, down-like fibre that grows from December to July from secondary skin follicles beneath coarse guard hair on some double-fleeced goats. It grows on some dairy goats, some Boer goats and a high percentage of New Zealand feral goats (Anon, 2011). Such goats are shorn before the down sheds (Lambert, 1990). Cashmere is considered to be the premium or ultimate animal fibre and is therefore not as affected by fashion as mohair is. Returns have been relatively constant for 80–90 years, with few of the major downturns in price suffered by mohair and wool (Anon, 2011).

The mean fibre diameter (MFD) of cashmere is 14–18 microns. Feral goats may yield 50 g of cashmere per animal annually. Through breeding and selecting the best animals, 300–400 g per year can be produced. Fibre returns reflect the yield of de-haired down, and there are specifications for minimum length (30 mm), fineness (maximum 18 microns MFD), diameter distribution, colour and lustre. Premiums are available for the finest fibre, no colour and no lustre (Anon, 2011).

Important cashmere production factors are highly heritable, so rapid genetic progress can be made through selection and planned breeding. However, MFD will increase rapidly with down weight (DW). Fibre testing is expensive; fibre length is a good indirect measure of DW. Feed levels have much less of an effect on production than with other products. Liveweight, fleece weight and fibre diameter do, however, tend to increase with an animal's age. In particular, fibre diameter will increase by approximately 1 micron from one to two years of age and DW by about 40–50 per cent. Pregnancy has an inverse relationship with cashmere production, reducing cashmere growth (Anon, 2011).

Cashgora

Cashgora fibre, 19–23 microns in diameter, is intermediate between cashmere and mohair fibre. It is obtained from Angora goats crossed with other goats. There is a limited, low-priced and variable export market for cashgora.

Shearing

Goats lack the insulation that sheep get from their layer of subcutaneous fat. The period following shearing is therefore risky if it coincides with low feed supply, cold and stormy weather or pregnancy. Measures must be taken before and after shearing to minimise stress and/or death for shorn does. In the 4–6 weeks prior to shearing, adequate pasture cover should be built up for the post-shearing period, and plans made to cope with stormy weather at shearing time. Does should not be shorn within three weeks of kidding, and neither should animals in poor condition. Shearing should end early in the day to allow animals to get to a familiar, sheltered paddock with adequate feed.

Angora goats are shorn twice annually and regrowth commences immediately. They are therefore not at as much risk as cashmere goats, which are shorn in winter (Lambert, 1990).

Flock of Angora goats.

Calendar of Angora goat farming operations

January	Shearing
February	Shearing
March	Mating — bucks put in with does
April	Mating ends
May	Culling of unwanted bucks
June	
July	Shearing
August	Shearing
September	Kidding
October	Docking and de-horning
November	
December	Weaning of kids

Meat production

The demand for New Zealand goat meat has always fluctuated. From 1928 to 1932, significant production levels were achieved before rising incomes (following the Great Depression) led to a fall in demand. From 1943 to 1946, large-scale exports of goat meat to the United Kingdom were undertaken to meet the wartime demand for meat, but following the end of the war and a return to normal conditions, demand for New Zealand goat meat once again fell to negligible levels (Sheppard & O'Donnell, 1979). In 2012, chilled or frozen goat meat was identified as the twenty-fifth most valuable export product of the year with a total value of US$4 million. However, at US$4.98 its price per kilogram was $0.48 less than in 2011 (MBIE & MPI, 2014b).

Goat meat has a significant worldwide market but a limited domestic market within New Zealand. International demand for processed goat meat has outstripped world supply in recent times, resulting in 2005 prices being double those achieved in 1990. Historical export markets for goat meat have been Trinidad and Tobago, Italy, Fiji and the Netherlands; in 1976–77, the total value of goat meat exports to such markets was just NZ$839,000, equivalent to NZ$0.99/kg. However, in 2008–09 the major export markets for New Zealand goat meat were the Caribbean (35 per cent), Africa (34 per cent) and North America (21 per cent). For the year ending September 2009, goat product exports (excluding dairy) totalled NZ$7.19 million, comprising $5.99 million for goat meat, $0.74 million for goat fibre and $0.46 million for goat skins.

Table 5.2 **Export kill and carcass weight for New Zealand goat meat**

Year	Export kill (thousands)	Export carcass weight (tonnes)
1997–98	147	11,613
1998–99	133	1523
1999–00	105	1179
2000–01	132	1506
2001–02	23	1264
2002–03	103	1219
2003–04	109	1279
2004–05	117	1321
2005–06	95	1059
2006–07	106	1148
2007–08	93	996
2008–09	107	1183

Source: Anon (2011).

Fluctuations in the quantity of meat exported from New Zealand are apparent from the data presented in Table 5.2. These fluctuations are however less extreme when compared to the mohair industry, suggesting that meat production may be a more stable source of income for farmers.

Breeds

The Boer goat comes from Southern Africa and is primarily a meat goat. Landcorp first imported Boer embryos into New Zealand from Zimbabwe in 1987, 1988 and 1989, but the animals did not become commercially available until 1993 when they were released from quarantine. A second company also imported embryos from Zimbabwe in 1987 and began a breeding programme; animals from this group were also released in 1993 (NZ Boer Goat Breeders Association, 2014).

For commercial meat production in New

Boer goats.

Zealand, Boer stud animals are crossed with Saanen does to provide more milk for kids and with cashmere does for thicker coats to protect against a colder climate than Boer goats are accustomed to in South Africa. Boer goats are large, with does weighing more than 50 kg and bucks up to 100 kg. They have high fertility, rapid growth rates and produce high-quality carcasses. Boer goats are suited for crossbreeding with other goat breeds to improve meat characteristics.

Under commercial conditions, pure-bred Boers can have growth rates in the range of 50–250 g/day up to 5–8 months of age, compared with 40–100 g/day for feral/cashmere/Angora goats (Anon, 2011). Under good management, Boer goats will reach carcass weights of 14–18 kg at eight months of age. Boer goats are quiet to handle and not likely to jump fences.

There is a market for goat carcasses in America for consumption by Muslims, Italians and Mexicans, and also one for some top-grade meat in Japan. Within New Zealand, at least 500 carcasses a week enter the Auckland market. Premium or top-grade meat comes from a young Boer or Boer cross animal that still has its kid teeth, has a carcass weight of up to 18 kg and is not much more than 12 months of age. As at 2014, returns for this product were $4.50/kg. In New Zealand, goat meat has been given the Heart Foundation Tick as a healthy, low-fat product with one-third of the fat content of New Zealand lamb meat.

The Kiko breed was developed in New Zealand by Garrick and Anne Batten of Nelson for meat production; the Māori word kiko means flesh or meat. The breed was established by crossing feral does with Nubian, Toggenburg and Saanen bucks, with further crossbreeding in the second and third generations. After four generations of selective breeding on the basis of survivability and growth rate in a hill-country environment, a dramatic improvement in liveweight and animal performance was achieved. The Kiko herd book was closed in 1986, meaning that all breeding animals now have to be selected from within the herd.

Kiko goats are approximately twice the weight of feral goats and are capable of producing good amounts of meat under a variety of New Zealand conditions. However, the Kiko breed has not become popular as a meat goat, with the Boer breed finding greater favour. Consequently, development of the Kiko ceased in about 1992. However, Kikos were exported to the United States in the 1990s, and at the beginning of 2005 there were more than 1600 Kiko goats registered with the International Kiko Goat Association in North America. A breeding programme has since developed the Kikonui (NZ Rare Breeds, 2012).

Farm goats

Goats mustered off farms in the North Island make up a significant proportion of the national goat kill. Premium prices of up to $4.50/kg (as at 2011) are possible until goats cut four teeth, by which time a carcass weight of 20.5 kg can be reached (Anon, 2011). A limitation to the New Zealand goat meat industry is that producers are restricted as to when they can kill animals, as many freezing works within the North Island only accept goats during the off-season. Killing and freezing charges are also often higher than for sheep (Sheppard & O'Donnell, 1979).

Goatskin

A by-product from all goat industries (but particularly the meat industry) is goatskin. In 1979, approximately 80 per cent of the skins produced were exported in a salted state to Europe, the United Kingdom and Korea. The remaining 20 per cent were used by a New Zealand firm for processing into leather. The most highly demanded skin is that of young kids up to one week old. Goatskin may be used for a wide variety of products, such as kid gloves, shoe uppers and liners, book bindings, bags, clothing and chamois. Feral goat skins tend to be too small for use in upholstery and are often of a poor quality as a result of lice, mange and tick damage.

Weed control

On many farms, goats are used for weed control. Such goats are generally run in association with sheep and cattle and provide meat and skins on culling. They also have positive implications for pasture quality and clover availability for other stock (Sheppard & O'Donnell, 1979), and can provide cost savings from reduced herbicide application. Goats are agile, and are able to eat vegetation up to 2 metres off the ground through their standing and climbing techniques. They have a narrow muzzle and a mobile upper lip, allowing them to penetrate into the browse and to selectively graze the high-quality parts of the forage. Their ability to control weeds depends on the type, sex, age and number of goats available and the particular weed species.

The stocking rate required for successful weed control is dependent on the pest plant species, the available goats and the system to be used. On a mixed farm, a stocking rate of 250–300 kilograms of liveweight per hectare (kg LW/ha) (set stocked) on mixed scrub weeds should stop weeds from spreading. To reduce weed numbers, strategic stocking with larger mobs at 2000 kg LW/ha for short periods will be needed (Anon, 2011). An alternative method of looking at stocking rate is in terms of ewe equivalents (stock units, SU). It is estimated that one feral milking doe and kid is equal to 0.8–1.0 SU, and studies indicate that a stocking rate of 8 SU per hectare of solid scrub will meet the feed requirements of goats.

However, further research has indicated that a stocking rate of three times maintenance is required to reduce an area with 40 per cent scrub cover by 50 per cent in one year (Sheppard & O'Donnell, 1979). This stocking rate is not sustainable and will be detrimental to goat health and survival, as goats do not cope well under heavy grazing/browsing pressure (Sheppard & O'Donnell, 1979).

Goats used for weed control purposes should be above 18 months of age, have already reached their mature liveweight and have had the target weed included in their diet in the past. Wethers (castrated males) are preferable. The system should allow goats to graze both on and off the target weed. Goat-proof fencing is critical.

Goats have been used successfully to control gorse, mānuka, sweet briar, broom, thistle, bracken fern, barberry, rushes, sedges and blackberry (Lambert, 1990). They will also strategically control undergrowth in pine forestry blocks where leaders are greater than 2 metres high and the trunk is protected by a skirt of unpruned branches. Following pruning, they can be used permanently among pines when trunk circumference is greater than 50 cm and the bark is corky (Anon, 2011).

Goats can have a strong impact on thistle numbers. Thistles are often not consumed until they have flowered, at which stage the goats graze down the flowering stem and into the base of the plant. This greatly reduces seed production, breaking the life cycle of this annual/biennial species. Woody species such as gorse and broom are also readily controlled by goat grazing. Rushes are not highly preferred by goats, but with long-term stocking or short-term, high-density stocking they can be successfully controlled (Lambert, 1990).

A figure of $10,000 of savings annually on an average-sized hill farm from reduced herbicide purchase and application, for gorse control in particular, has been commonly quoted in relation to using goats for weed control. Goats also increase the grazing area for sheep and cattle, maintain pasture quality and increase the legume (white clover) content in pastures, all of which have direct implications for the performance of other stock (Lambert, 1990).

Opportunities exist to make money from farm feral goats. For little or no input apart from an occasional mustering, feral goats can provide an extra return to a farm business. A case study farm of 2000 hectares with approximately 2000 feral goats has estimated carcass returns of $25–$30 per feral goat if caught and processed for meat (Meat & Wool New Zealand, 2008). Animals targeted for culling are all unwanted bucks (better bucks can be kept for mating purposes), all older does and wethers, and those goats with an apparent liveweight of 30 kg (giving a carcass weight of 12 kg). Feral goats are also valuable due to their ability to control a wide range of weeds, as discussed above. It is estimated that on the case study farm, the effect of feral goats on weed control equates to a saving of $40,000 per year, or $20/ha in the case of this 2000 ha property (Meat & Wool New Zealand, 2008).

Animal welfare

Goats are very social animals, living in hierarchically organised groups (Schino, 1998). It is therefore very important to introduce new goats into a herd gradually and to monitor signs of stress. Goats are vocal animals and will let the handler know when they are in distress or in pain. New Zealand has a code of welfare for goats, which specifies what is considered to be optimal animal welfare and how this may be achieved for goats farmed under conditions specific to New Zealand (National Animal Welfare Advisory Committee, 2012).

Saanen and Saanen cross dairy goats being entertained outdoors.

Animal health

Diseases

The principal diseases of dairy goats as reported by milking goat farmers in New Zealand are shown in Figure 5.1.

Caseous lymphadenitis (CLA or 'cheesy gland') is common to sheep and goats. It is caused by *Corynebacterium* infection of the skin and results in external abscesses. It is best managed by vaccination. Johne's disease is a wasting disease that is prevalent in many ruminants; the actual prevalence of Johne's in goat herds is unknown. In goats it can be managed by a combination of vaccination, culling and strict biosecurity to prevent introduction of the disease.

The main goat-specific disease is caprine arthritis and encephalitis (CAE), an insidious viral disease that occurs worldwide. The main clinical sign is swollen knees, but it also causes production losses and reduced immunity. The virus is spread between goats through infected milk or colostrum; it is killed by heating the milk or colostrum to 56°C for one hour (Adams et al., 1983). CAE can also be managed by preventing kids from drinking colostrum or milk from infected does. Goats are also tested for CAE antibodies to see if they have been exposed to the virus; this allows farmers to identify and progressively remove any goat that might be a carrier of the CAE virus.

Somatic cell count (SCC) is used by the dairy industry as a measure of milk quality. The count is elevated in the milk of goats with udder infection, but there is a large overlap between SCC levels in milk from infected or uninfected glands. It is thus difficult to establish an SCC limit that reliably differentiates infected and uninfected glands in goats. The SCC in goat milk is also affected by stress, oestrus and stage of lactation (Lerondelle et al., 1992; McDougall & Voermans, 2002; Wilson et al., 1995) and remains elevated once infection is cured (McDougall et al., 2010). Growing bacteria from milk is the only way to confirm an infection in goats. A survey of farms determined that most udder infections in dairy goats were from coagulase-negative *Staphylococci* (13.4 per cent) and *Corynebacterium* species (7.3 per cent; McDougall et al., 2014).

Foot health and lameness is an issue for dairy

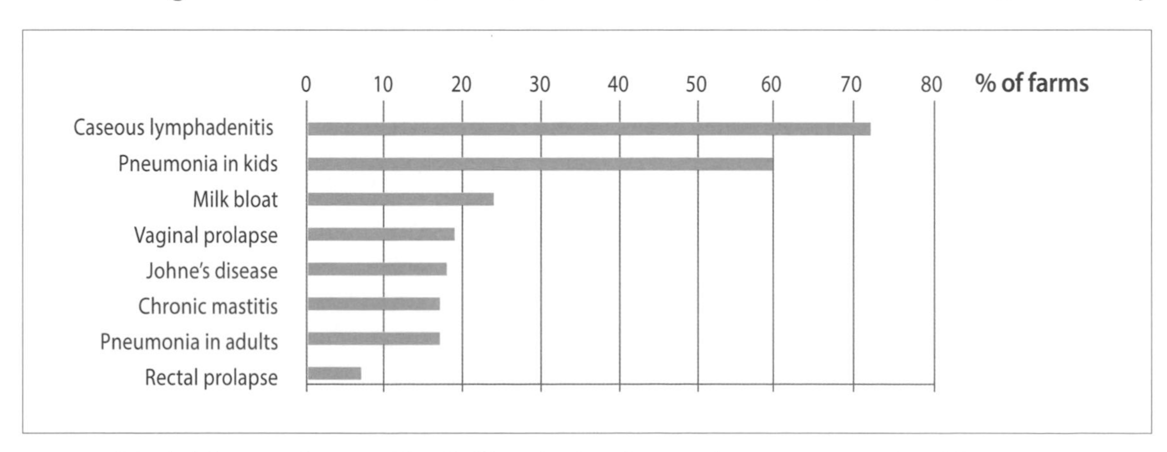

Figure 5.1 Principal diseases of goats as identified by Dairy Goat Co-operative goat farmers in 2014/15.

goats, as the hoof wall grows continuously and wet conditions increase the risk of foot diseases due to bacterial infections. To manage this rapid growth and prevent lameness, a goat's hooves are usually trimmed 2–4 times a year. Feeding is also important: protein fed to goats encourages rapid growth of the hooves, and feeding high levels of carbohydrates can lead to laminitis.

Veterinary medicines for milking goats

Only a small number of veterinary medicines are licensed for use in goats. Although some drugs can be used 'off-label' by veterinarians for animals in their care, the veterinarian must take responsibility for the risk of failure of the product or problems with residues in milk. This is a difficult area, as there are few studies on the metabolism and excretion of drugs in goats. It is not always possible to predict withholding periods by comparing goats with other species (Smith & Sherman, 2009).

Vaccinations are available for Johne's disease, CLA, toxoplasmosis abortion, campylobacteriosis abortion, clostridial diseases (i.e. pulpy kidney, blackleg, tetanus, and septicaemia) and leptospirosis. Not all vaccines have been proven to be effective in goats, and not all farmers use them. Treatment options for internal parasites in milking goats are limited due to resistance and to the milk withholding periods of anthelmintics (Watson, 1994). The majority of dairy goats are now farmed off paddock and fed parasite-clean pasture to avoid this problem.

Growing kids

Newborn dairy goat kids must receive colostrum within the first 6–8 hours of life, to protect them from harmful infections. This requirement must be balanced against the need to prevent the spread of CAE disease from dam to kid. Cow colostrum is often used as an alternative to goat colostrum; alternatively, synthetic colostrum may be used.

Dairy goat kids are usually raised in pens (20–25 kids per pen). The number of kids per pen and the size of the pen must be managed to avoid smothering, as the kids like to lie tightly together and sometimes on top of each other. The growth targets for kids are between 100 and 250 g/day. The average weaning age is between 10 and 12 weeks, but can be anywhere between eight and 15 weeks. The average weight at weaning is between 17 and 20 kg. The weight at mating (at around nine months) is generally between 60 and 75 per cent of mature doe weight. Hay and/or meal is fed to kids from about two weeks of age; at weaning, kids should be eating at least 250 g of meal twice daily.

It is common practice to disbud dairy goat kids between three and 10 days of age. Disbudding prevents the growth of horns so that the goats do not become a danger to other goats in the herd or to themselves. Horned goats have a tendency to bully and can injure other goats (Smith & Sherman, 1994). At present the most common method of disbudding is to apply heat to the buds to destroy the cells from which horns will grow. Research is also being carried out to find alternative means and explore how anaesthetics or analgesia can be used to minimise distress.

Surplus kids

Due to the high number of kids born per milking doe, there are always more kids than are required for replacement milking does or breeding bucks. Some kids are raised for the cabrito market; cabrito is a cut of meat from goat kids around

Dairy kid goats being reared indoors.

4–8 weeks old, i.e. before weaning. This is not an option for kids that are underweight or sick. For euthanasing kids that are not to be reared, a captive bolt method has recently been introduced by the DGC as a more humane method than blunt force trauma. The captive bolt method provides the user with a very consistent means of ensuring immediate insensibility and death. The carcasses are then processed for skins and meat by-products.

Goats can be milked for several seasons without being dried off (Linzell, 1973); thus, another option to reduce the number of kids born each year is to only mate and dry off a small proportion of the milking goats in the herd in any year.

Conclusion

The New Zealand goat industry as a whole, and as individual industries, has had a turbulent history in terms of product value, export quantities and stock numbers. Goat farming in New Zealand has a chequered history, typical of many alternative livestock farming ventures. While currently there is high interest in goat milk products and hence the economic outlook for dairy goat farming is good, it will always remain a niche industry within New Zealand due to the higher fixed costs of farming goats. Farming milking goats can be problematic — partly due to the nature of the animals, but also because there is a lack of scientifically proven information on best farming practices. Many goat farming methods have been developed empirically by farmers, who then pass on their knowledge to other farmers. The dairy goat industry is not yet large enough to warrant the attention of farm advisors, a dedicated farm support industry, or goat-specific veterinary medicines. As a result, the industry is characterised by a 'make do' attitude. It can be expected that the goat industry in New Zealand will continue to be volatile, due to its reliance on overseas markets and due to available stock numbers, and will always be a niche rather than a substantial domestic industry.

References

Adams, D.S., Klevjer-Anderson, P., et al. (1983). Transmission and control of caprine arthritis–encephalitis virus. *American Journal of Veterinary Research*, 44, 1670–75.

Anon. (2011). Goat management and feeding. In G. Trafford & S. Trafford (Eds), *Farm technical manual*. Christchurch: CCH New Zealand.

Ceballos, L.S., Morales, E.R., et al. (2009). Composition of goat and cow milk produced under similar conditions and analyzed by identical methodology. *Journal of Food Composition and Analysis*, 22, 322–29.

Legarto, J., & Leclerc M.-C. (Eds). (2011). L'alimentation pratique des chèvres laitières (The practical guide to feeding dairy goats); pp. 101–09, 115, 123. Paris: Institut de l'Élevage (French Livestock Institute).

Lambert, M.G. (1990). Status of goat farming in New Zealand. *N.Z. Goat Research*, 59–64.

Lerondelle, C., Richard, Y., & Issartial, J. (1992). Factors affecting somatic cell counts in goat milk. *Small Ruminant Research*, 8, 129–39.

Linzell, J.L. (1973). Innate seasonal oscillations in the rate of milk secretion in goats. *Journal of Physiology*, 230, 225–33.

Martin, P., Szymanowska, M., et al. (2002). The impact of genetic polymorphisms on the protein composition of ruminant milks. *Reproduction and Nutrition Development*, 42, 433–59.

McDougall, S., Malcolm, D., & Prosser, C.G. (2014). Prevalence and incidence of mastitis in lactating dairy goats. *New Zealand Veterinary Journal*, 62, 136–45.

McDougall, S., Supré, K., De Vliegher, S., Haesebrouck, F., Hussein, H., Clausen, L., & Prosser, C. (2010). Diagnosis and treatment of subclinical mastitis in early lactation in dairy goats. *Journal of Dairy Science*, 93, 4710–21.

McDougall, S., & Voermans, M. (2002). Influence of estrus on somatic cell count in dairy goats. *Journal of Dairy Science*, 85, 378–83.

MBIE & MPI. (2014a). iFAB 2013 Dairy Review. Report prepared by CORIOLIS as part of the Food and Beverage Information Project. Wellington: MBIE. Accessed at www.mbie.govt.nz/info-services/sectors-industries/food-beverage/documents-image-library/Dairy%20sector%20review%202013%20-PDF%201.8%20MB.pdf.

MBIE & MPI. (2014b). iFAB 2013 Meat Review. Report prepared by CORIOLIS as part of the Food and Beverage Information Project. Wellington: MBIE. Accessed at www.mbie.govt.nz/info-services/sectors-industries/food-beverage/documents-image-library/Meat%20sector%20review%202013%20-PDF%201.6%20MB.pdf.

Meat & Wool New Zealand. (2008). Making money from feral goats. Goat monitoring project, information sheet no. 5. Wellington: Meat & Wool New Zealand. Accessed at www.maxa.maf.govt.nz/sff/about-projects/search/04-059/making-money-from-feral-goats_.pdf.

Mohair Producers NZ Ltd. (2016a). Home page — latest price update for fleece. Accessed at www.mohairproducers.co.nz.

Mohair Producers NZ Ltd. (2016b). Farming/Characteristics/Resistance to soiling. Accessed at www.mohairproducers.co.nz/principal-characteristics/.

National Animal Welfare Advisory Committee. (2012). Animal Welfare (Goats) Code of Welfare. Wellington: New Zealand Government.

New Zealand Boer Goat Breeders Association. (2014). About Boer goats. Accessed at www.nzbgba.co.nz/boergoats.

New Zealand Rare Breeds. (2012). Breeds/Kiko goats. Accessed at www.rarebreeds.co.nz/kiko.html.

Parkes, J.P. (2004). Feral goats. In C.M. King (Ed.), *The handbook of New Zealand mammals* (pp. 374–92). Oxford: Oxford University Press.

Robertson, K., Symes., W., & Garnham, M. (2015). Carbon footprint of dairy goat milk production in New Zealand. *Journal of Dairy Science*, 98, 4279–93.

Sheppard, R.L., & O'Donnell, D.K. (1979). A review of the New Zealand goat industry. Discussion paper 42. Christchurch: Agricultural Economics Research Unit, Lincoln College. Accessed at https://researcharchive.lincoln.ac.nz/bitstream/handle/10182/999/aeru_dp_42.pdf?sequence=1 (December 2015).

Schino, G. (1998). Reconciliation in domestic goats. *Behaviour*, 135, 343–56.

Silanikove, N., Leitner, G., Merin, U., & Prosser, C.G. (2010). Recent advances in exploiting goat's milk: quality, safety and production aspects. *Small Ruminant Research*, 89, 110–24.

Smith, M.C., & Sherman, D.M. (1994). Fundamentals of goat practice. In C. Cann, S. Hunsberger, R. Lukens & M. Denardo (Eds), *Goat medicine* (pp. 1–15). Lippincott Williams and Wilkins.

Smith, M.C. & Sherman, D.M. (2009). Appendix A. Formulary of some drugs used in goats and suggested dosages. In M.C. Smith & D.M. Sherman, *Goat medicine*, 2nd edition (pp. 807–14). Wiley-Blackwell.

Solis-Ramirez, J., Lopez-Villalobos, N., & Blair, H. (2011). Dairy goat production systems in Waikato, New Zealand. *Proceedings of the New Zealand Society of Animal Production*, 71, 86–91.

Solis-Ramirez, J., Lopez-Villalobos, N., & Blair, H. (2012). Economic values for New Zealand dairy goats. *Proceedings of the New Zealand Society of Animal Production*, 72, 166–68.

Watson, T. (1994). Anthelmintic resistance in the New Zealand animal production systems. *Proceedings of the New Zealand Society of Animal Production*, 54, 1–4.

Wilson, D.J., Stewart, K.N., & Sears, P.M. (1995). Effects of stage of lactation, production, parity and season on somatic cell counts in infected and uninfected dairy goats. *Small Ruminant Research*, 16, 165–69.

Chapter 6

Pig Production

Patrick C.H. Morel

Chapter 6

Pig Production

Patrick C.H. Morel

Institute of Veterinary, Animal and Biomedical Sciences
Massey University, Palmerston North

Introduction

Pigs were domesticated in Neolithic times and since then have lived closely with people, often sharing housing with them. For most of this time, pigs roamed around rural communities and towns feeding on food wastes and, probably, human excrement. In the countryside they may have been supervised in woods and forests, eating roots, nuts and berries. In the eighteenth and nineteenth centuries, more modern pig production systems developed. Pig breeds were produced, and pigs were housed and fed grains plus kitchen wastes. Intensive pig farming systems developed in the United States in the twentieth century and have been adopted by many countries. Pigs are farmed to be killed for meat. Pork, bacon and ham are popular meats worldwide, except for Muslims and Jews who consider pig meat unclean and do not eat it.

In New Zealand in the 1950s, the pig industry was very much allied to the dairy industry. Small piggeries were built on dairy farms, and the pigs were fed skim milk produced on the farm. Pig production tended to be inefficient and was of secondary interest to most dairy farmers. Pig production was seasonal, following milk production. When whole-milk collection by dairy companies started in the 1960s, the supply of skim milk as an on-farm by-product ended, and many dairy farmers stopped keeping pigs. Between 1966 and 1970, the number of farms with pigs declined by 40 per cent and the size of the national sow herd declined by 20 per cent.

The New Zealand Pork Industry Board, known as NZPork, is the main organisation servicing the pork industry. NZPork is currently funded solely through a levy on all pigs slaughtered in registered premises ($4–$5 per pig, revised annually). Sixty per cent of the levies are used for marketing purposes (e.g. Trim Pork and NZPork advertising campaigns). The function of NZPork is to promote and organise an efficient pork-producing industry in New Zealand.

In the 1970s more-specialised pig farms developed, and following a large increase in productivity from 1970 to 1985 the number of pigs marketed per sow per year stalled for many years until recently, when it increased again. These farms fed grain and by-products from various food industries including milk processing plants and breweries.

Intensive pig-production systems are characterised by the following features:

- Farms are on small acreages of land.
- Stock is housed throughout life in environmentally controlled buildings.
- Stocking density is high and large numbers of animals are housed under one roof.
- Feeding and manure removal are mechanised.
- Stock is fed expensive concentrate diets.
- High capital investment in stock, buildings and equipment is required.
- A businesslike approach to management is needed.
- Systems lend themselves to vertical integration.
- Farms are under scrutiny from animal welfare organisations.

Pigs are monogastric (single-stomached) and omnivorous. They are usually fed high-quality diets based on grain, as opposed to the high-fibre roughage diet typically fed to ruminants. Because they are omnivorous, however, pigs can also be fed food waste. Many pig producers obtain out-of-date baking products and restaurant waste (swill) to feed to pigs as a supplement to grain-based meals or pellets.

Outdoor and indoor piggeries in New Zealand are operated in a similar fashion to piggeries overseas. At a large piggery, all operations are organised on a weekly basis in that in virtually every week of the year sows are mated, piglets are born, piglets are weaned and grower (finished) pigs are sent to slaughter.

On a conventional indoor pig farm producing pigs for slaughter, piglets are born and remain with their mother for 3–4 weeks. They are then weaned and grouped together for growing. Most will be finished and sent to slaughter as porkers at about 115 days of age at 70 kg, or grown further and killed at about 4.5 months of age as baconers at 90 kg. Gestation is about 114 days, weaning takes place at 21–28 days of age, and rebreeding

Commercially crossbred New Zealand sow and feeding piglets outdoors. **Previous:** Hampshire pigs.

10 days later means that each reproductive cycle is 152 days. A sow can thus have up to 2.5 litters per year, with each litter being 10 or 11 piglets.

Replacement breeding females (gilts) may be sourced from breeding companies and be mated at 7–8 months of age (130 kg). Pregnant gilts and sows are put into a pregnant sow (dry sow) pen with other gilts or sows. A couple of weeks before they farrow (give birth), they will be moved into a maternity shed and usually put into a farrowing crate in which they will give birth. After 3–4 weeks of suckling, the piglets are weaned and the sow is put into a mating shed. She might be put into a stall until she is bred and then, once pregnant, into a pregnant sow pen. Mating may be by artificial insemination (AI) or be natural.

In outdoor piggeries, pregnant sows may share a paddock with other sows. In late pregnancy they will be put into a small paddock by themselves with a shed (ark) in which they will farrow. Weaned piglets may be reared outdoors, or partially outdoors, or in sheds like in a conventional indoor system.

Because of the high costs and the small profit margins, pig farmers need to adapt rapidly to technological changes if they wish to maintain their profitability.

Table 6.1 **Basic facts for growing pigs and female replacement stock**

Stage	Name	Housing	Age (day)	LW (kg)	Diet	FCR
Growing pigs						
Birth	Piglet	Farrowing area	1	1.4	Sow milk Creep feed	1.2–1.4
Weaning	Weaner	Weaner deck	21–28	6–8	Weaner feed	1.5–1.8
Transfer	Grower	Growing shed	63–70	25–30	Grower feed	2.2–2.4
	Finisher	Growing shed	95–105	60–70	Finisher feed	2.4–2.6
Slaughter	Porker	Growing shed	115	60–65	Finisher	2.6–2.8
Slaughter	Baconer	Growing shed	147	85–95	Finisher	2.8–3.0
Replacement stock						**FI (kg/day)**
Selection	Gilt	Gilts shed	150	85–95	Dry sow	3–3.5
Mating		Mating area	220	130	Dry sow	2.3
Pregnancy weeks 1–4		Mating area			Dry sow	2.5
Pregnancy weeks 5–15		Dry sow area			Dry sow	2.5
Pregnancy week 16+		Farrowing area			Dry sow	2.0
Lactation week 1	Sow	Farrowing area			Lactating sow	3–3.5
Lactation weeks 2–4	Sow	Farrowing area			Lactating sow	up to 6–10
Weaning	Sow	Mating area			Dry sow	3.5 (flushing)
Mating	Sow	Mating area			Dry sow	2.3

FCR = feed conversion ratio (kg feed per kg liveweight gain); FI = feed intake (kg feed per day); LW = liveweight.

Glossary

Dry sow: a non-lactating sow.

Farrow: give birth to piglets.

Farrowing crate: narrow crate holding a sow from before farrowing to weaning.

FCR: feed conversion ratio (kilograms of feed per kilograms of liveweight gain).

FI: feed intake (kilograms of feed per day).

Finisher: pigs that are generally above 70 kg liveweight, until they are sold or retained for breeding.

Gilt: a young female pig, selected for reproductive purposes, before she has had a litter of piglets.

Grower: pigs generally between 30 and 70 kg liveweight.

Lactating sow: a sow that has given birth and is producing milk to feed her piglets.

Mated gilt: a young female pig that has been mated but has not yet had a first litter.

Sow: an adult female pig that has had one or more litters.

Weaner: a pig after it has been weaned from the sow up to until approximately 30 kg liveweight.

The New Zealand pork industry

The pig farming industry in New Zealand is small by international standards. Today, there are around 100 commercial pig farms in New Zealand, with an average herd size of 300 sows. The majority are held indoors, but approximately 40–45 per cent are kept outdoors in free-range or free-farmed operations. There are around 26,000 sows in New Zealand and 590,000 pigs are slaughtered annually, producing 42,100 tonnes of meat slaughter weight. From 1990 to 2016, average pig carcass weight increased from 56.74 kg to 71.35 kg. This reflects the tendency for the proportion of pigs over 60 kg liveweight at slaughter to increase, and to now represent about 95 per cent of total finished pigs.

In 2016, the value of pig production in New Zealand was around $160 million at the farm gate or $420 million as retail (sold to consumers). Despite being relatively small by international standards, the pig industry is an important domestic industry and about 21 kg of pork is consumed per capita per year (Figure 6.1). Only a small amount of pig meat is exported, mainly to the Pacific Islands (192 tonnes in 2016).

Although the amount of pig meat imported in the 1980s was minimal, it has steadily increased since then. Today, New Zealand imports about 55 per cent of its needs, mainly for processing (Figure 6.1). Until recently, processed pig meat only could be imported, but in 2014 the government allowed the importation of fresh pork in cuts of less than 3 kg. Pig meat is imported from Canada, the United States, Denmark, Finland, Australia and Sweden. The pork industry argued against the importation of fresh pork because of concerns about the possible introduction of

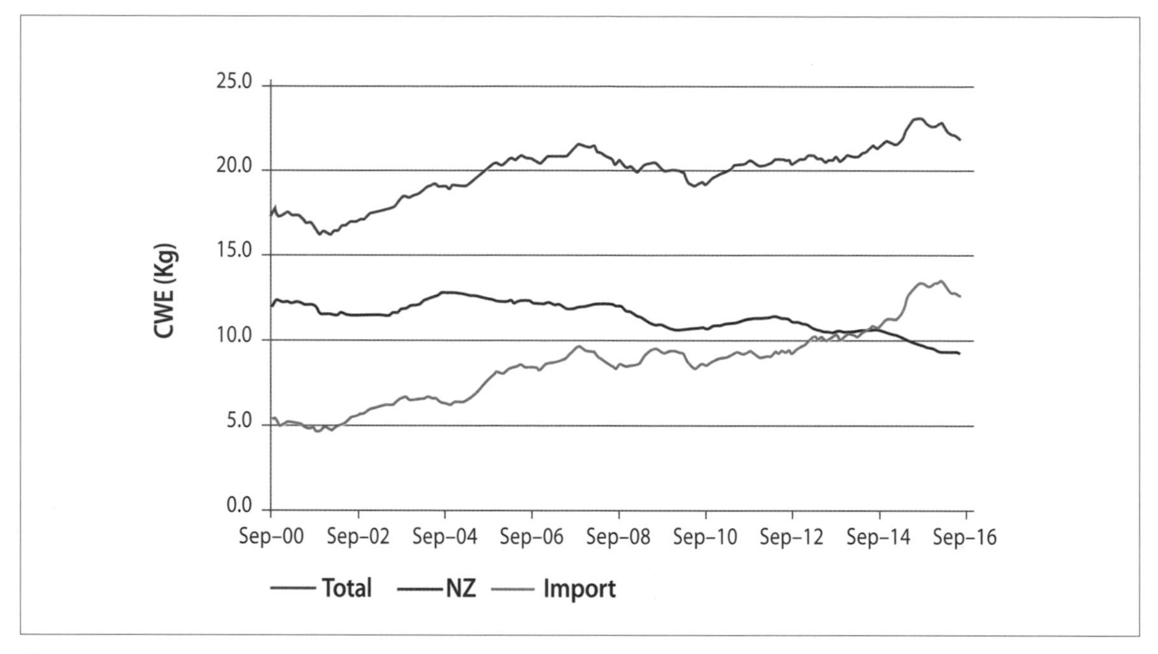

Figure 6.1 Pork consumption per capita in New Zealand as kg carcass weight equivalent (CWE) for total, imported and locally produced pork. Collated from various sources.

pig diseases not found in New Zealand, such as porcine reproductive respiratory syndrome (PRRS), porcine coronaviruses (transmissible gastroenteritis [TGE], porcine epidemic diarrhoea [PED]), systemic salmonellosis (caused by *Salmonella enterica* serovar Choleraesuis), and — most importantly for New Zealand as a whole — foot and mouth disease (FMD).

Currently, pig farmers face several problems including high feed prices, increased imports, new animal welfare and waste disposal requirements, and fluctuating pig meat prices. The importation of fresh and processed pig meat from countries where farming is subsidised has put pressure on pig farming in New Zealand. This, combined with concerns about pig welfare which has resulted in the elimination of sow stalls, has made pig farming more difficult. The increase in pig meat imports weakens local control over how the pig meat we eat is produced. Pig welfare standards are poorer in some of the countries from which we source pig meat than in New Zealand.

The price paid per kilogram of pork carcass to the farmer is generally higher in the North Island than in the South Island. The relationship between the baconer price paid per kg of carcass to the farmer in New Zealand and the cost including freight (CIF) of imported pork cuts is shown in Figure 6.2. This illustrates that the price in New Zealand is directly related to the international market price.

The consumption per capita of pig meat in New Zealand is increasing slowly in comparison with poultry meat and there is strong competition in price and marketing from beef and lamb.

In 2005, data on key production parameters were obtained from around 30 per cent of New Zealand's commercial pig farms and compared with data from countries exporting pork to New Zealand. Overall, the level of production in New Zealand was equal to that of most of its

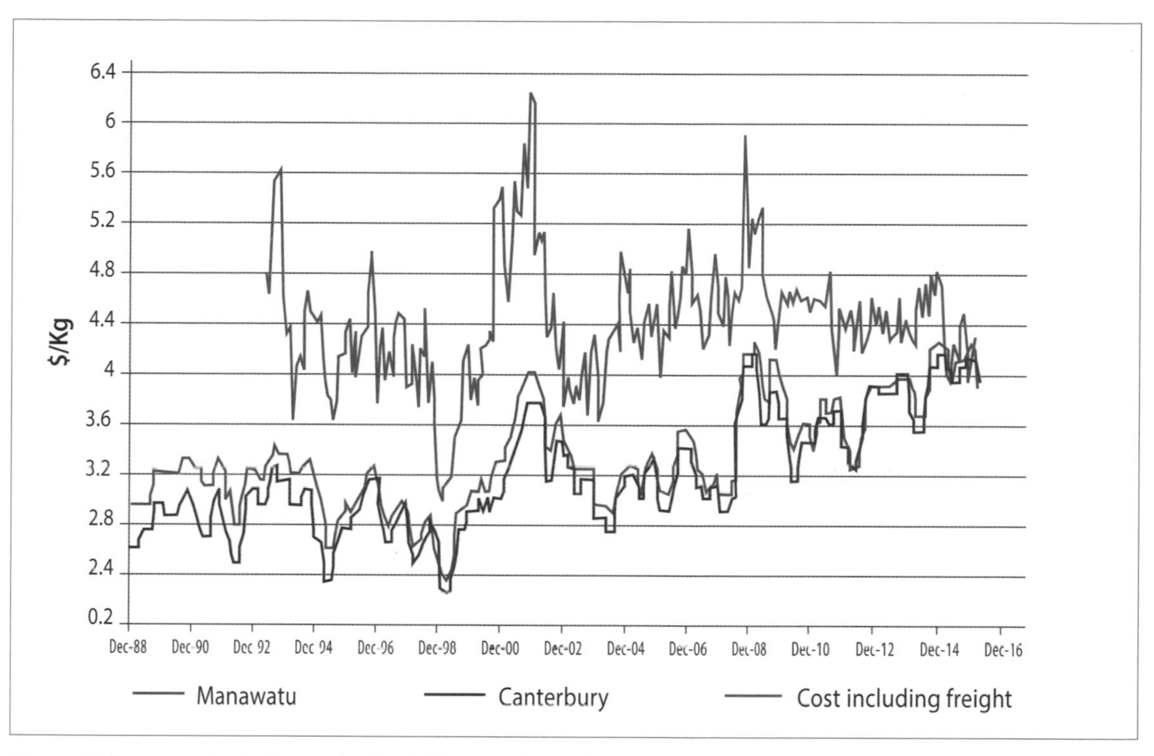

Figure 6.2 Baconer price in the North Island (Manawatu) and the South Island (Canterbury), and cost including freight for imported pig meat between 1988 and 2016. Collated from various sources.

key trading partners. The lack of international competitiveness of the New Zealand pig industry is therefore not due to poor performance, but instead mainly to higher feed costs, smaller farm sizes and thus higher production costs (Davidson, 2005).

To meet these challenges and stay in business, New Zealand pig producers have to either reduce production costs or develop high-quality products that fetch a premium price. The pig industry is technologically advanced, has an aggressive marketing policy, operates marketing forecasting models to adjust production levels, and encourages the production of a high-quality lean product produced with a high standard of welfare (e.g. Trim Pork, PigCare™).

The number of pig farmers who will survive these challenging times is unknown, but the challenges they face are substantial.

Pig farming shifted from the mainly dairy-farming districts to the grain-producing districts in the 1970s. The proportion of pigs produced in the South Island has increased in the past 30 years: it was around 65–70 per cent in 2016 compared with 30 per cent in 1980. Pig feed is based on grain (barley, wheat, corn), animal protein (fish meal, meat and bone meal, blood) and plant protein (soya beans, peas). Feed prices are lower in the South Island, where most New Zealand grain is produced.

In decreasing order of importance, the main factors influencing the profitability of a pig farm, expressed in terms of gross margin, are price for the finished pig, feed costs (Figure 6.3), feed conversion ratio, number of pigs sold per sow per year, carcass grading profile and pig growth

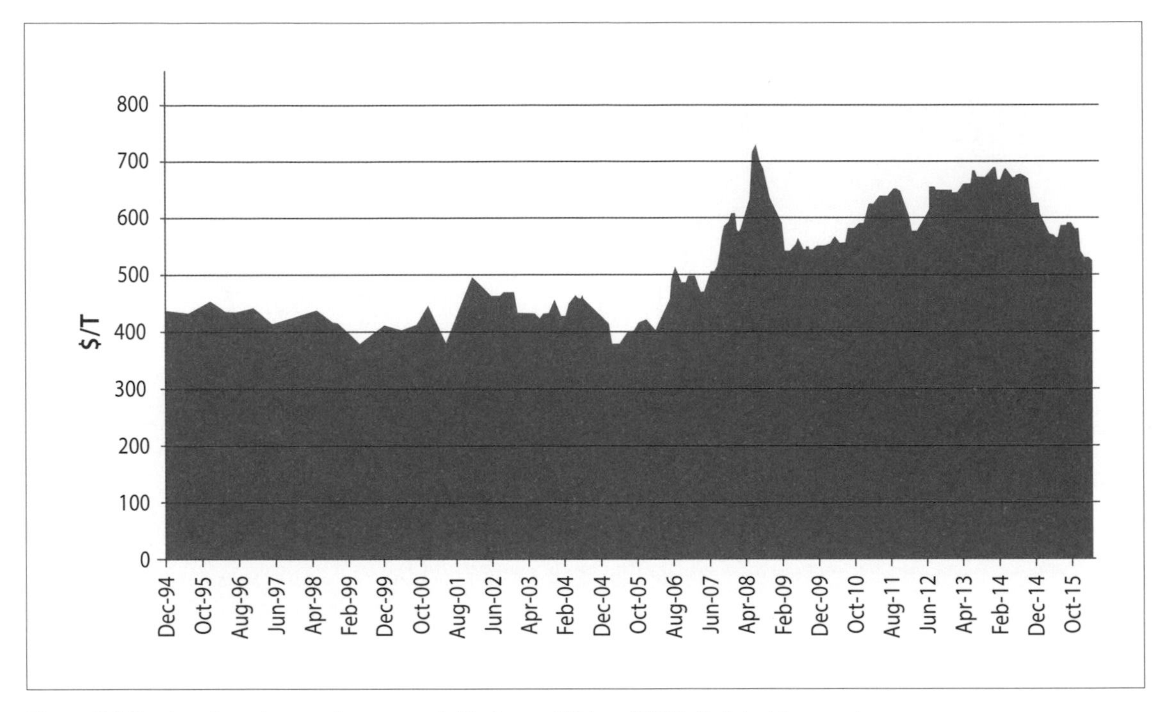

Figure 6.3 Feed cost per tonne of a grower diet between 1994 and 2016. Collated from various sources.

rate. Individual pig farmers have a limited influence on the price schedule and feed prices, which are both driven by the international market. Today, with modern genotypes and appropriate nutrition, it is not difficult to achieve the maximum price within the carcass grading profile. Pig growth rate has the smallest effect on profitability. Pig farmers must therefore concentrate their efforts on increasing the number of pigs sold per sow per year and improving the feed conversion ratio. Pig farmers use genetics to select appropriate pig breeds and crosses to improve these characteristics. They also adjust diets to maximise growth and yield per sow and per piglet born.

In a pig carcass, the thickness of backfat is closely related to the percentage of lean meat and is used as a criterion for the selection of leaner animals. Backfat thickness is also the basis of the carcass grading and payment system. The aim of any carcass classification scheme is to provide a basis for payment to the producer according to the financial value of the carcass to the consumer. The price per kg slaughter weight (price schedule) is determined as a function of the carcass weight (CW) and the backfat thickness. The price varies in a stepwise (non-continuous) fashion. It is common to have a 35–40 cent drop per kg carcass weight when backfat thickness increases from 12 mm to 13 mm.

Pig carcasses are valued according to their leanness. The percentage lean is usually estimated from backfat, and sometimes from muscle thickness measurements made after slaughter. Measurements are taken using a ruler (not so accurate), intrascope (accurate but slow), Fat-O-Meter (FOM) or Hennessy grading probe (HGP) (accurate and fast, but expensive). In New Zealand, carcasses are valued on a single backfat measurement taken 6.5 cm off the mid-line

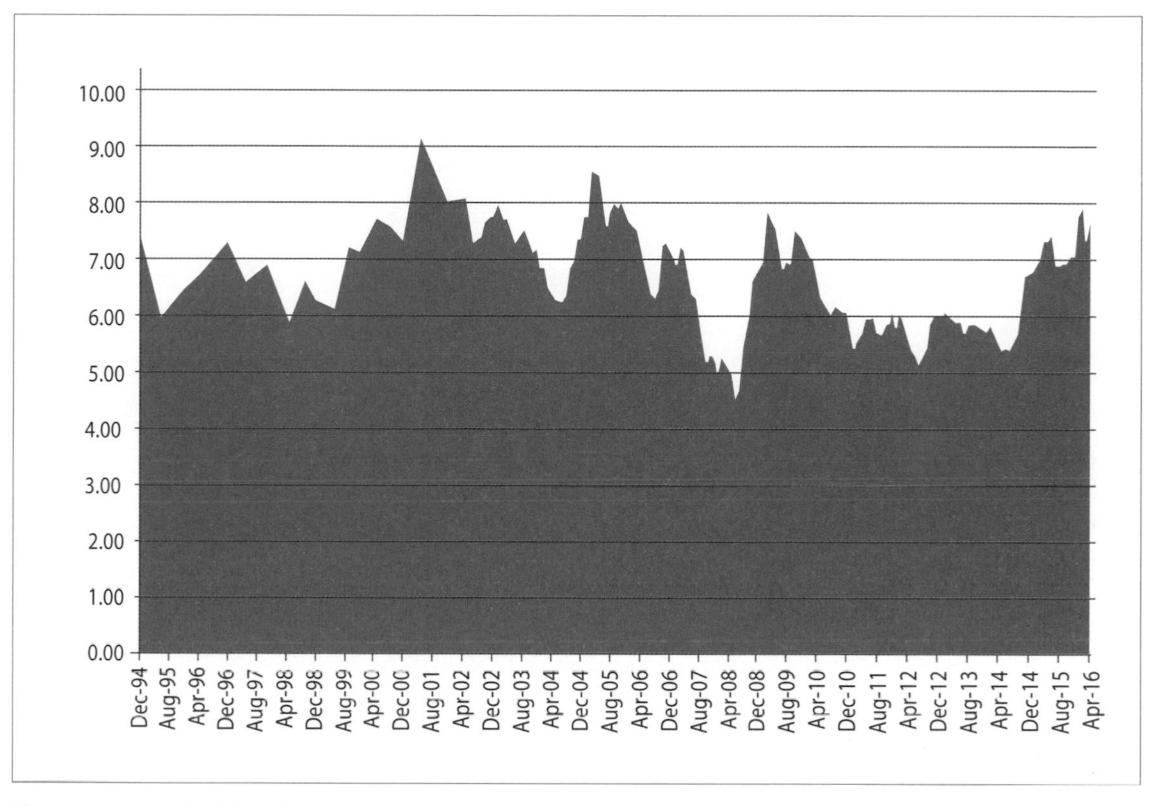

Figure 6.4 Price schedule to feed cost ratio between 1994 and 2016. Collated from various sources.

at the level of the last rib (P2), using an optical probe (HGP) or intrascope.

There are two main indicators of efficiency for pig meat production on pig farms:

- **Feed to meat ratio** is related to the efficiency of production. This ratio is usually calculated over one year as the total feed used divided by the total kg carcass produced. Typical values in New Zealand are between 3.5 and 5, which are comparable with those in Australia, North America and Europe. Pigs in New Zealand are produced efficiently.
- **Price schedule to feed price ratio**, which is an indicator of the economic environment in which pig production is taking place. At average levels of on-farm efficiency, it is usually considered that to allow for a reasonable return on capital after depreciation, the ratio should exceed 7. The changes in this ratio between December 1994 and March 2016 are illustrated in Figure 6.4. Since 2005, the ratio has been nearly consistently below 7; the economic pressure brought about by this has resulted in a reduction in the total number of pig farms in New Zealand.

Production systems

Housing and temperature

Pig health and performance are influenced by the environment, and the housing system should provide environmental conditions that stay within the optimal range for each class of pig.

Animals try to maintain a constant body temperature (39°C). The air temperature below which food is oxidised to produce extra heat to maintain this body temperature is known as the lower critical temperature (LCT); for farrowing sows this is 13°C, for newborn pigs 25–30°C, and for growing pigs 16–21°C. High and very low temperatures are undesirable. Under hot conditions, pigs keep cool and try to stay under the upper critical temperature (UCT) by spreading out the body, seeking shade, drinking water, wallowing and increasing the respiration rate. In pigs, sweat glands are almost non-functional. Pigs have a relatively narrow zone of thermoneutrality between the LCT and the UCT (Figure 6.5).

The LCT depends on the following factors: feeding intensity, liveweight, group size, air temperature, air motion and bedding material. An increase in the feeding intensity will decrease the LCT as high energy intake increases heat production. Fat animals have a lower LCT, because fat cover is a good insulator; therefore, leaner pigs need to be kept warmer. Pigs huddle to reduce heat loss. Air motion increases heat loss through convection and elevates the LCT. In hot conditions, increasing air velocity will alleviate heat stress (convective cooling). Bedding and floor type influences heat loss by conduction, with higher LCTs found on concrete slats and

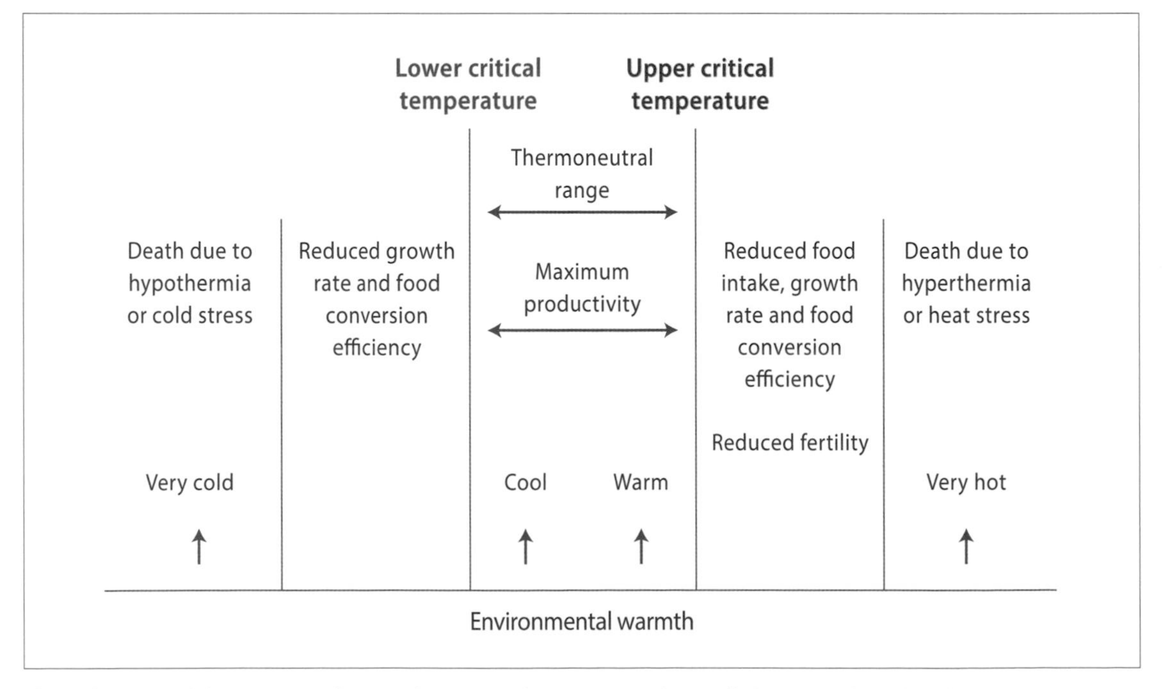

Figure 6.5 General descriptions of pig productivity and environmental warmth. Source: Holmes (1982).

lower LCTs on straw. As body weight increases, the LCT decreases (from 30°C in piglets at birth to 10°C for 100 kg pigs).

The temperature requirements of sows are between 13°C and 18°C, and for piglets between 30°C at birth and 24°C at three weeks of age. The farrowing accommodation (maternity ward) should thus provide two different environments, one for the sows and one for the piglets.

Piglets have little fat or hair for insulation and a large surface area in proportion to their body mass. The body temperature of piglets born in a cold environment declines rapidly due to their inability to maintain their heat balance, thus compromising their survival. There is a high piglet mortality rate between birth and weaning, 7–20 per cent depending on the housing system, and the majority of deaths occur during the first few days of life. The majority of these deaths are not due to diseases but to starvation, overlaying (crushing) and weakness at birth. Piglets having a low birth weight, less than 1 kg, are most at risk.

While piglets need a high environmental temperature to increase their chances of survival, a high temperature in the farrowing house causes the sow to become restless and may result in overlaying and piglet crushing. Moreover, the sow's appetite will also decline, which causes a decrease in milk yield and thus piglet body weights. Furthermore, the sow will lose more weight during lactation. Two thermal environments within the farrowing house or hut must therefore be created, with the piglets' nest provided with supplementary heat. An infra-red heat bulb and reflector are traditional, but gas heating and floor warming are also used.

When adequately fed and in good body condition, adult sows and boars do not need elaborate housing requirements, with temperature requirements between 13°C and 18°C. Pigs live happily out of doors in temperate climates when a simple shelter to protect them from the elements is provided. With complete indoor housing of sows during pregnancy, ambient temperature comes under human control. In summer a major problem is keeping temperatures down; in winter the removal of excess moisture from buildings becomes a concern, as air change in pig houses is reduced to conserve heat.

Piglets on flat deck.

Relative humidity

This factor is only likely to be a problem in artificially heated buildings and is more of an issue in winter. In pig buildings there is usually no direct control over humidity levels. Provided that the ambient temperature and ventilation rates are correct, relative humidity (RH) will settle to a satisfactory level (65–80 per cent). When RH is higher than 80 per cent, the resultant condensation leads to undesirable effects on respiratory health. When RH is lower than 65 per cent, a drying-out of skin and the mucous membranes of the upper respiratory tract occurs, leading to irritation, reduction in appetite and loss of productivity.

Ventilation

Ventilation is needed to remove noxious gases (carbon dioxide, ammonia), dust particles (which act as transport agents for disease and irritate the respiratory passages), moisture produced by stock (through respiration, and evaporation of urine and faeces) and to control indoor temperature and humidity. Ventilation must prevent excess heat leaving the building in winter, while being sufficient to keep temperatures down within the comfort zone in summer. Higher, forced ventilation may be needed in summer than in winter. An increased air velocity adds stress in cold conditions through heat loss by convection, but in hot conditions alleviates heat stress.

Lighting regimen

Natural light is involved in the synthesis of vitamin D, but this vitamin can now be supplied in the diet. It is generally agreed that the duration and the intensity of lighting have no effect on grower pig performance. Weaners are less active and show fewer vices with low-intensity lighting. With breeding stock, lighting effects are unclear; however, there is some evidence that the lighting regimen post-weaning influences the onset of oestrus (possibly a cause of summer infertility).

Housing systems

Empty and pregnant sow housing

Outdoors

Free-range and outdoor breeding systems are extensive systems where sows and boars are kept outdoors in open paddocks. There is a mating area and a gestation area. Climatic conditions and soil type must be suitable for this type of farming. Usually, free-draining soils are preferred. However, with this type of soil, unsuitable conditions can lead to excessive nutrient leaching into the water table and nearby streams and rivers. The critical factor in minimising leaching is maintaining a good grass cover all year round. The South Island, particularly the Canterbury region, is well suited for this type of pig farming.

Compared with indoor farming, outdoor pig farming is more labour-intensive and has higher land costs, more health problems (parasites, pneumonia, arthritis) and higher feed costs (as more feed is needed to maintain body condition), but has lower building costs.

The main reasons for culling sows (sow wastage) are reproductive failure (32 per cent) and lameness problems (32 per cent). Reproductive failures are associated with sows being in poor condition (low fat reserves). New strains of pigs with a high percentage lean are, therefore, not well suited for outdoor systems. More robust genotypes should be used in the dam lines.

Indoor stalls

In traditional intensive indoor systems, sows were kept in individual sow stalls (gestation stalls) from weaning until farrowing. This system allows the accurate observation of animals, individual feeding, mechanised feeding and waste removal, and an easily controlled climate. However, a high standard of management is needed, and insulation and ventilation requirements entail high initial costs. One of the major welfare issues in sow management is the inability of the sows to move freely when in sow stalls. Since December 2015, sows in New Zealand can be kept in stalls only between weaning and mating, which is around one week. From the week after mating until a few days before farrowing the sows must be kept in pens.

Indoor group housing

Buildings used to house pregnant sows are usually low-cost, eco-shelter-type buildings with straw or sawdust as bedding. The advantages of keeping sows in groups are a greater freedom of movement; individual choice between available micro-environments; more opportunity for social interaction; improved cardiovascular fitness, muscle weight and bone strength; decreased morbidity; and less abnormal behaviour.

Sow aggression can be a problem in group housing. Sows have a hierarchical social structure that is achieved through fighting and aggression. This is particularly obvious at mixing (after weaning/mating) and at feeding time. The level of aggression in a group often escalates pre-farrowing and there is an increase in skin lesions and vulva biting. Another challenge with group housing is the difficulty of feeding individual sows their correct feed allowance to prevent either obesity or thinness. To overcome this difficulty, different feeding systems have been developed. These include floor feeding, which is spreading feed over a large area; feeding stations and the slow distribution of feed (trickle feeding); the use of individual feeding stalls (bails); electronic transponders to trigger feed release; and pipeline feeding into troughs.

A wide variety of non-confinement group housing systems is used to accommodate pregnant sows in New Zealand. They can be classified into static and dynamic systems.

- In a static system, sows are kept in small groups (10–20). The groups comprise all of the sows mated at the same time (on a weekly basis); the sows may also be sorted according to size and body condition, and the gilts are kept separately. No new sows are added to any group and sows which have had to be removed are not added back. Thus, all the sows in the group are at the same stage of gestation.
- In a dynamic system, sows are kept in large groups (100–200). Each week, all of the newly mated sows are added to the large group and sows close to farrowing are removed. Thus, all stages of pregnancy are present in a dynamic group. The sows also tend to form sub-groups; for example, all those mated in the same week tend to stay together.

Pregnant sows in fully slatted pen.

Sow and litter housing

Outdoors

Sows about to farrow are held in individual, warm, dry, draught-free and clean individual huts with straw bedding. There are no manure disposal problems. Relatively low capital outlay is needed for the housing and equipment, but adequate acreage for rotation, free-draining soils and a moderate climate are needed. A higher feed intake also occurs. This is a simple system; the scale of the operation and the type of sow are the major factors in success. Providing an environment within the hut with the high temperatures required by piglets is critical for their survival.

There are various types of hut. The main aspects to be considered are size, presence of guard rails, flooring, door location (centre or side), skids and hooks for moving huts, straw availability, insulation and ventilation. Construction materials are not critical, but huts must be robust. Staff safety aspects should also be taken into account — it is important to keep the sow away from staff when they are handling piglets.

Indoors

Farrowing crates have been adopted worldwide since the late 1960s; they have many advantages. They ease sow management in large herds and minimise the crushing of piglets. They allow the piglets access to the sow's underline, have a self-cleaning floor and a separate heated creep area for the piglets, and contain a feeder and a

Sow and piglets in farrowing crate.

drinker. Sows farrowing at the same time are kept in the same room, which allows better disease control. This is an all in, all out system, with cleaning and disinfection of the rooms between farrowing groups. As each sow is confined in a crate, the staff have easy and safe access both to the sows during farrowing and to the piglets.

In April 2016, following a review of the farrowing crate system, the National Animal Welfare Advisory Committee (NAWAC) concluded that currently there are no alternatives to farrowing crates in term of piglet welfare. Piglet mortality is at its lowest in farrowing crates and at its highest in outdoor farrowing systems. The use during lactation of farrowing pens, where sows have more space and can turn around freely, has recently been investigated on a commercial farm in New Zealand. Farrowing pens with temporary crating up to four days after birth were compared with traditional farrowing crates. Pre-weaning piglet mortality was higher in the pen system (10.2 per cent) than in the crate system (6.1 per cent), and the total piglets weaned per litter was lower in the pen system (10.5) than in the crate system (10.8) (Chidgey et al., 2015).

Weaner pig housing

After weaning, piglets may be housed outdoors in huts, verandah-type housing or flat-deck housing.

- Huts are a low-cost shelter with an outdoor run. Straw or sawdust bedding is used directly on the ground or on a concrete floor. The number of weaner pigs per hut can vary from 20 to 200, depending on the system.
- Verandah-type housing is suitable for piglets at 4–5 weeks of age. There are no environment controls, so a high level of insulation is needed. Warm kennels are provided with access to an outdoor run, there is a slatted dunging area and a floor feeding system operates.
- Flat-deck housing is suited to piglets from weaning to 9–10 weeks of age. The buildings have insulation and controlled ventilation and temperature (22–28°C), with cages over a pit that is flushed down daily. The floor is mesh, plastic or iron slats. Nine to 15 piglets are held in each pen and they are fed ad libitum. This system is costly to erect and operate.

Grower finishing houses (25–100 kg)

Outdoors

Pigs are kept in an open paddock with access to low-cost shelter. The same constraints apply as for outdoors systems for empty and gestating sows.

Indoors

The buildings have a high degree of structural insulation. Ventilation is controlled and stocking rates are high (varying from 20 pigs per pen to 100-plus). Feeding may be ad libitum floor feeding. Slatted floors help with cleanliness. Mechanical feeding, like wet feeding systems, may be used. The main problems are outbreaks of disease and vices such as tail-biting. Cheaper buildings, such as straw or plastic houses (eco-shelters), are also used for growers.

Housing for breeding boars

Breeding boars are usually housed separately but close to weaned breeding sows. They may be in stalls or in yards. Boars are moved through the breeding sows daily to identify sows in oestrus.

Waste management

Pig effluent can include faeces, urine, cleaning water, rainwater, soil, bedding, waste feed and spilt drinking water. The management of the effluent produced by housed indoor pigs is a significant part of farm management.

The different processes involved in managing effluents before application to land are shown in Figure 6.6. These processes can include:

- Storage in pits under the shed in a 'pull plug' system, flushed to sumps and then either irrigated directly to land or discharged to anaerobic/aerobic ponds, sometimes via a solids separation unit.
- Flushing directly to sumps and irrigation to land.
- Flushing directly to anaerobic/aerobic ponds.
- Flushing through a solids separation unit followed by discharge to ponds or irrigation to land.
- Material scraped 'dry' and slurry spread to land.
- Deep-litter systems where spent bedding is either spread directly on land or composted in piles and then spread on land or sold on.
- Separated solids either spread directly on land, or stored and composted and then spread on land or sold on.
- Direct discharge to land for outdoor pigs.

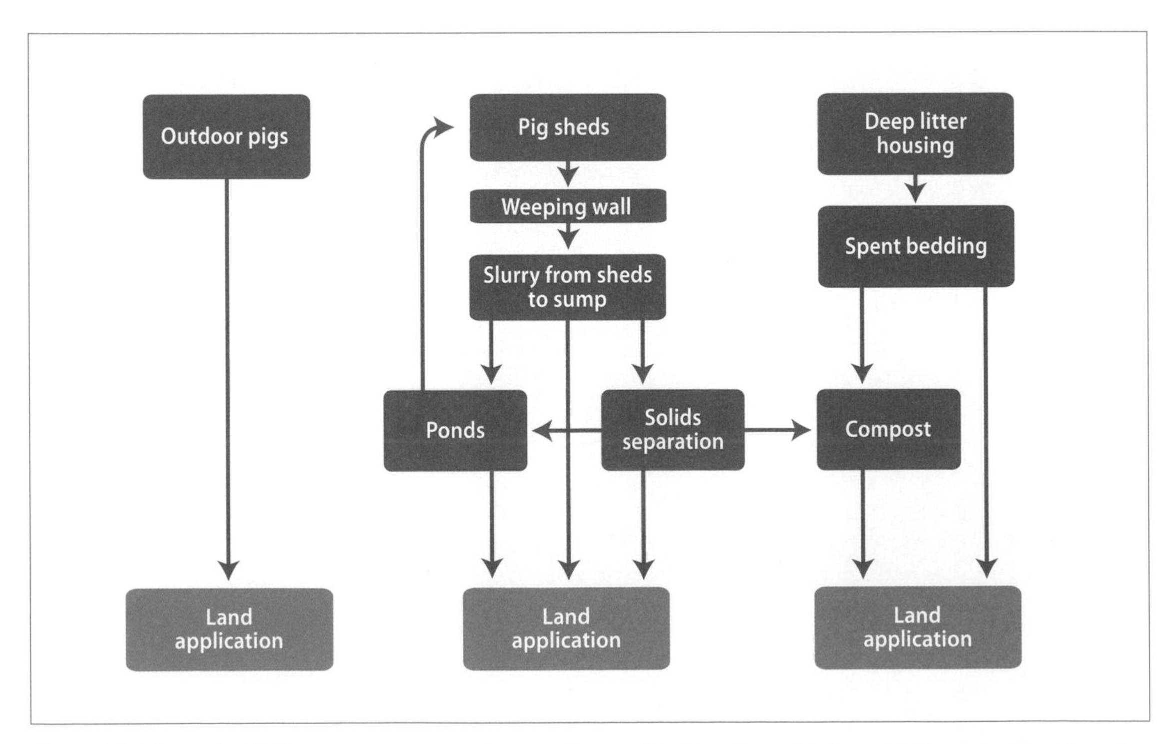

Figure 6.6 Schematic of effluent management processes. Source: Good practice guide for nutrient management in pork production, NZPork (2014).

The nutrients in the effluent can lead to land and water quality management issues for piggeries. The main issues that affect water quality are nitrogen, sediment, phosphorus and pathogens. The Resource Management Act (RMA), passed in 1991, is New Zealand's principal legislation for environmental management. This Act of Parliament promotes the sustainable management of natural and physical resources such as land, air and water. Across New Zealand, farmers as well as regional councils are asking for increased levels of information about nutrient leaching rates from primary production activities. Regional councils, through their regional plans, are signifying that their preferred (and in some cases required) method of analysing and reporting this information is through the software package OVERSEER Nutrient Budgets (http://overseer.org.nz/).

Sows and piglets outdoors with feeding unit.

Genetics

There are three ways to use genetics to improve productivity in a pig herd. First, all of the breeding pigs on a commercial farm could be replaced regularly by superior breeds or strains. Second, superior animals could be identified and used for breeding within a purebred population (nucleus herd). Third, crossbreeding between different breeds could use the benefits of hybrid vigour (heterosis) to increase productivity on multiplier and commercial farms.

The genetic structure of a population is usually represented as a triangle with animals of higher genetic merit towards the top. In the pig industry, the pyramid structure consists of the nucleus herd (selection within a purebred population), multiplier herds (producing crossbred replacement stocks), and commercial producers obtaining breeding boars and sows from multipliers for the production of a slaughter generation (Figure 6.7).

Nucleus herd

The nucleus herd is composed of pigs of the specific breeds used on farms in New Zealand. New Zealand has two major pig breeding companies: PIC NZ Ltd (www.picnz.co.nz) and Animal Breeding Services (ABS; www.abreeds.co.nz). Both import genetic material in the form of semen from the United States or Norway; the pure-breed pig genotypes available in New Zealand have thus been selected overseas using selection indices developed in these countries. Each company operates its own genetic improvement programme.

A selection index allows several traits to be selected for and improved at the same time. In constructing a selection index, the relative economic values, heritabilities, phenotypic and genetic variances for each trait, and phenotypic and genetic correlation between traits all have to be known (Tables 6.2 and 6.3). In the

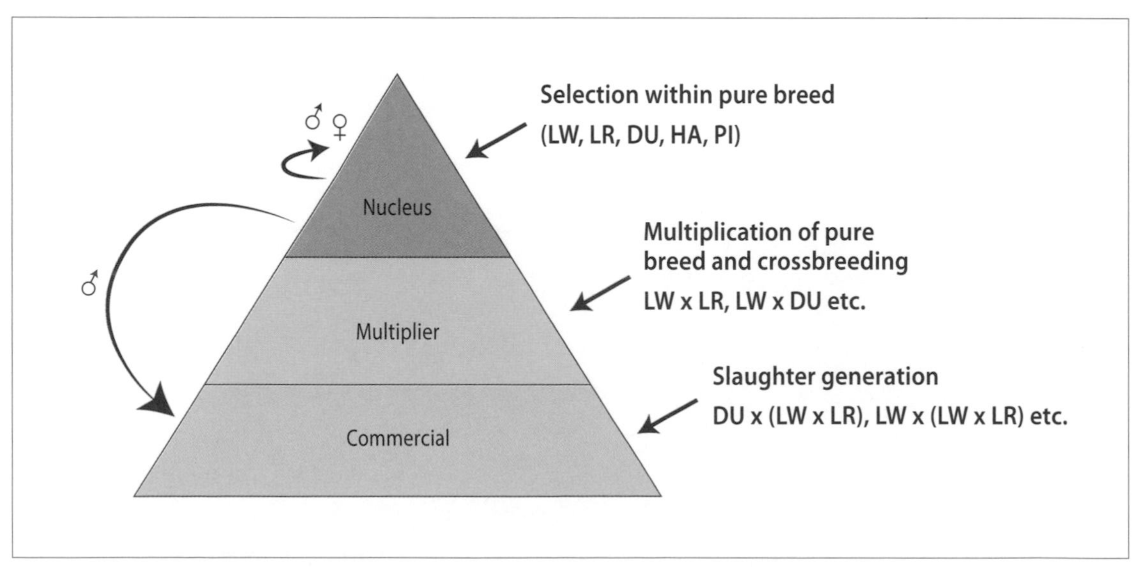

Figure 6.7 Genetic structure of the pig industry. DU = Duroc; HA = Hampshire; LR = Landrace; LW = Large White; PI = Piétrain.

Main commercial pig breeds

Large White

Large Whites are distinguished by their picturesque bearing, erect ears, slightly dished faces, white colour, pink skin and long deep side. They originated in Yorkshire, England. They have excellent reproduction, growth and carcass performance.

Landrace

Landrace swine are white in colour with heavy, drooped ears, and are long and muscular. They have excellent reproduction, growth and carcass performance. However, this breed may have some meat-quality problems.

Duroc

The Duroc is a red breed that varies in colour from a very light gold to a very dark red. They have good reproduction, growth and carcass performance and an excellent meat quality. When used as terminal sires, progeny are white-skinned.

Hampshire

The Hampshire has a shiny black hair coat with a white belt over the shoulder, and has a reputation for good grazing and foraging abilities. They have reasonable reproduction performance, good growth and carcass performance and a good meat quality.

Piétrain

This breed comes from Belgium. Piétrains are white with black spots. They are of medium size and extremely lean.

international pig industry, breeding programmes are highly sophisticated and the selection indices used in the nucleus herds will include a combination of many possible traits. Usually sire lines (boars) will be selected on the basis of average daily weight gain, feed conversion ratio, percentage lean (or backfat thickness), meat quality, feed intake and intramuscular fat, whereas dam lines (sows) will be selected for litter size, piglet survival and robustness.

The flow of genetic improvement is illustrated in Figure 6.7. For example, boars from the sire line may be used to breed sows in the nucleus herd as part of the selection programme in the multiplier farm to produce crossbred sows, and again in a commercial herd (using AI) to produce the slaughter generation.

Multiplier herd

In a multiplier herd, crossbreeding between different breeds or strains takes place, mainly on the female (dam) side. The dam lines are cross-mated to produce a first-cross sow, known as an F1, which is available to the commercial farmer. The principal objective of mating individuals from different populations (breeds, strains) is to take advantage of non-additive genetic effects (dominant, overdominant and epistatic effects), which are the basis of the heterosis phenomenon (Table 6.4). Heterosis, or hybrid vigour, is defined as the difference in performance between the crossbred and the average of the parental breeds. The largest heterosis values are observed for reproductive performances; this means that crossbreeding is mainly used to improve reproduction traits.

Other traits, like growth rate, feed conversion ratio, carcass quality and meat quality, which have low heterosis but high heritability (Table 6.2), must be improved by direct selection in the pure breeds. On commercial farms, Duroc or 50 per cent Duroc boars are often used as terminal sires to improve growth and carcass quality of the slaughter generation.

Table 6.2 **Heritability values (0–100%) for traits commonly used in pig selection programmes**

Trait	Level	Range (%)	Mean (%)
Litter size at birth	low	5–15	10
Litter size at weaning	low	5–15	10
Weight at weaning	low	5–15	10
Growth rate	high	45–55	50
Feed conversion ratio	medium	25–40	35
Backfat thickness	high	40–60	50
Carcass leanness	high	40–60	50
Meat quality	medium	20–40	25
Feed intake	medium	10–75	35
Intramuscular fat	high	30–80	50
Leg weakness	medium	20–40	30
Fatty acid composition	high	40–70	55

Table 6.3 **Range of phenotypic (above diagonal) and genotypic (below diagonal) correlations between traits calculated for Large White, Landrace and Piétrain breeds under ad libitum feeding**

		FCR	ADG	% lean	FI	MQ	IMF
FCR	min		−0.40	−0.02	0.15	0.08	0.06
	max		−0.55	−0.56	0.61	0.18	0.18
ADG	min	−0.37		−0.15	0.43	−0.05	0.03
	max	−0.60		0.05	0.79	0.05	0.09
% LEAN	min	−0.04	−0.37		−0.12	−0.25	−0.11
	max	−0.74	0.16		−0.56	−0.06	−0.21
FI	min	0.08	0.48	−0.25		−0.05	0.15
	max	0.55	0.84	−0.72		0.11	0.18
MQ	min	−0.34	−0.53	−0.42	−0.26		−0.05
	max	0.34	0.16	0.00	0.06		−0.07
IMF	min	0.03	0.28	−0.30	0.41	0.16	
	max	0.13	0.40	0.00	0.50	0.19	

% lean = percentage of lean cuts in the carcass; ADG = average daily gain; FCR = feed conversion ratio; FI = voluntary feed intake; IMF = percentage of intramuscular fat; MQ = meat quality score. Sources: Blum (1983), Krieter (1986), Morel et al. (1988), Schmidt (1988) and Wäfler (1982).

Commercial herd

By using F1 (first-cross) sows, a commercial farmer producing pigs for slaughter utilises the benefits of heterosis mainly in the reproductive ability and robustness of the dam line. These crossbred sows are mated to purebred or cross-bred terminal sires to produce pigs, all of which go to slaughter.

Currently, replacement rates in the sow herd are around 50 per cent. This replacement rate is high, indicating that sow longevity is not optimal. This situation has resulted from modern genotypes having been selected for a high lean meat content, increased litter size and high milk production. Body fat reserves below 10 per cent are associated with reproductive failure. Modern genotypes have minimal body fat reserves at the first mating and lose body fat reserves after each farrowing; thus, they reach the critical 10 per cent of body fat reserves earlier than did the

Table 6.4 **Average individual and maternal heterosis as a percentage of mid-parent value**

Trait	Heterosis (%)	
	Individual	**Maternal**
Age at first oestrus	8	–
Litter size at birth	2	6
Litter size at weaning	6	9
Survival rate birth–weaning	4	2
Piglet weight at weaning	5	0
Litter weight at weaning	12	10
Growth rate after weaning	6	0
Age at 100 kg	5	0
Feed conversion ratio	2	0
Body composition	0	0
Meat quality	0	0

older, fatter sow types.

In choosing a replacement policy, the commercial farmer should take into account the following factors:

- genetic progress wanted (the level of genetic lag between breeder and producer)
- health status (buying in increases health risks)
- cost of stock

Buying replacement gilts may reduce the genetic lag but increase health risks. There are three options for a commercial farmer. These are:

- Home replacement gilts: boars are bought from nucleus herds and sows are bred on farm. With this system there are low initial costs and health risks, but greater genetic lag and only two-thirds of the crossbred advantages with criss-crossing.
- Buying in all replacements: normally, first-cross females and purebred boars are bought. This results in a small genetic lag and has maximum heterosis, but there are higher costs and health risks.
- Own multiplication unit: purebred sows and boars are bought from a nucleus herd to produce the farm's crossbred females. There are fewer health risks and a small genetic lag, but not all stock is crossbred and the system is only viable financially in large units (300+ sows).

In New Zealand, commercial farms usually obtain all their replacement stock (gilts and boars) from one source, to preserve their high health status. In addition, some larger farms have the entire breeding structure on farm, with small nucleus and multiplier herds, to produce their own female replacements. Commercial crossbred sows are mated via AI with boar semen from terminal sire lines purchased from the breeding companies.

Free-range, outdoor pig production.

Sow and piglets raised in an outdoor housing system.

Definitions in genetics

Phenotypic variance: the variance in the values observed when a trait is measured in individual animals.

Genetic variance: that part of the phenotypic variance that is caused by differences in the genetic make-up of the animals.

Phenotypic and genetic correlation between traits: the calculated correlation between two traits measured in a series of animals is called the phenotypic correlation (*rp*). This can be divided into an environmental component (environmental correlation, *re*) and a genetic component (genetic correlation, *rg*).

Heritability: the heritability (*h*2) of a trait indicates the extent to which differences in a trait are passed from one generation to another. It is a measure of the proportion of genetic additive variance in the phenotypic variance.

Heterosis (hybrid vigour): the difference in performance between the crossbred and the average of the parental breeds.

Breeding value: describes the merit of an individual's genes in terms of their values when passed to the next generation.

Economic value: an estimate of the financial value of a unit change in a trait.

Selection index: usually calculated by simply multiplying the breeding value by the economic value of each trait and summing the products.

Criss-crossing: a two-breed rotational cross-breeding system which involves crossing each female with the breed that was not its sire.

Reproduction

Male

Boars normally attain sexual maturity around 5.5 to 6 months of age. The total ejaculate of sperm is 150–300 ml, with a sperm concentration of 200–300 million sperm/ml. An ejaculate includes three fractions: a pre-sperm fraction (5–20 per cent clear fluid), a sperm-rich fraction (30–50 per cent) and a gel fraction (40–60 per cent gelatinous). At service, the sperm is delivered into the cervix. The gel fraction prevents backward leakage of the sperm through the cervix.

Copulation is a long process and can take up to 30 minutes. Eighty per cent of the sperm are in the first third of the ejaculate. The minimum duration of a successful mating is at least three minutes. On average, one boar is kept for every 15–20 females. When the number of services per week exceeds six, the litter size usually becomes reduced. A long rest period (longer than one month) will also increase the return rate (sows coming back into oestrus) and reduce the litter size.

Boars are usually used for two years before being replaced. Due to their size, older boars are used to mate older sows and younger boars are mated with gilts and young sows.

Female

Gilts (young female pigs) attain sexual maturity or puberty at 4.5 to 6.5 months (around 90–100 kg), with some small deviation due to nutritional status. The onset of puberty is influenced more by age than by weight. Crossbred gilts attain puberty 10–14 days earlier than purebreds. Gilts in contact with boars will also attain puberty earlier than gilts that are isolated from boars. The first exposure to the boars should occur between 160 and 170 days of age. Young boars (6.5 months), though fertile and capable of successfully serving the gilts, are not capable of stimulating the onset of puberty. A better stimulation is obtained when several boars are used in rotation. Transport stress or mixing stress will also stimulate the onset of puberty.

Once puberty is attained, pigs will show signs of oestrus every 21 days on average, and this lasts about 50–60 hours (shorter in gilts). The onset of oestrus is marked by swelling and reddening of the vulva. However, this occurs several days before the onset of 'true heat' and the shedding of the eggs. Ovulation occurs 18–36 hours after the beginning of heat. The number of ova shed varies from 10 to 25 and is greater in mature sows (15–25) than in gilts (7–16). There is a rapid rise in ovulation rate in the first two cycles after puberty.

Sows do not normally show signs of oestrus during lactation. Partial weaning of the litter to stimulate oestrus has been attempted, but this has not been successful. Oestrus normally occurs 5–10 days after weaning, depending on the body condition of the sow at weaning and on lactation length (weaning before piglets are three weeks old will delay oestrus). The gestation period ranges from 110–119 days. The mean is 114–115 days (three months, three weeks and three days). Optimal age at mating is around 220 days, with the first farrowing at approximately one year of age. Optimal lactation length is 3–4 weeks, and production of 2.2–2.45 litters per sow per year is possible.

Prenatal mortality of unfertilised ova

The conception rate (non-return rate) following natural mating is usually about 85 per cent. Failure to conceive is due either to mating at the wrong time, or to sow or boar problems (overwork, heat, nutritional and physiological stress, poor body condition). High returns may, however, be due to a loss of fertilised ova resulting from heat stress or disease. In a healthy, well-fed pig herd, the conception rate is primarily influenced by the timing of mating.

Prenatal mortality of fertilised ova

Sows may shed 10–25 ova, of which 90–95 per cent are fertilised. However, only 8–14 piglets per litter are carried to term, giving a 30–45 per cent loss of ova. Of this loss, 70–90 per cent occurs before day 25 (implantation) and 10–30 per cent after implantation. Early mortality is followed by reabsorption, later mortality by mummification and, later, expulsion. Disease (leptospirosis) or nutritional disorders may lead to death of the fetus followed by abortion.

Artificial insemination

The main limitations to using AI in pigs are the need for a high dosage rate of semen, the short storage life of fresh semen, and the low rate of sperm survival for frozen semen. Artificial insemination is widely used in New Zealand. It allows a pig farmer to use better genetic material over a larger number of sows, with minimal health risks. The results in the field are satisfactory with double inseminations. Farrowing rates are larger for AI combined with natural service and with pooled semen. Farrowing rates are also better with three services per oestrus rather than two or one.

Semen collection from the boar is mainly done using dummy sows. The training of boars to use dummy sows presents no major problems. Collections are made every 4–5 days. The amount needed is 50 ml of fresh undiluted semen — which is a large volume but necessary due to the tortuous female reproductive tract. One diluted ejaculate will give 20–30 services (200–300 million sperms /dose).

Fresh undiluted semen is viable up to a maximum of 24 hours. With normal diluent (at 15–20°C), semen can be stored for up to three days with fertility remaining normal; by five days, there is a drop in farrowing rate and litter size. New complex diluents have extend the viability of the semen up to 5–7 days.

Semen can be either collected by the farmer from a boar kept on farm or collected from boars in a nucleus herd by the breeding companies and sent by courier to the farm.

Sows and piglets in ark.

Nutrition

The pig is a monogastric, omnivorous animal. It is classified as having a simple digestive tract. The large intestine has a large population of microbes that carry out fermentation. The pig does not possess enzymes for degrading complex non-starch polysaccharides (cellulose, hemicellulose, lignin, pectin), although the large intestinal bacteria are capable of some limited degradation. Short-chain fatty acids produced by the bacteria make a small contribution to the animal's overall energy supplies. Vitamins are not synthesised by the bacteria to a large extent. Thus, in modern intensive pig production the animals are supplied with all key nutrients in a readily digestible form (energy, protein/amino acids, essential fatty acids, minerals, vitamins and water).

Energy

Energy is not a structural nutrient per se, but rather the 'fuel' for production. Energy can be supplied from starches and simple sugars, fats and oils, fibre, amino acids and alcohol. In New Zealand, pig diets are specified in terms of digestible energy (DE) content, in units of megajoules (MJ). In other countries, metabolisable energy (ME) or net energy (NE) may be used. The apparent digestible energy (ADE or DE) is defined as the gross energy of feed less the gross energy of faeces.

Dietary energy is used for the following processes:

- Maintenance, i.e. maintaining essential body functions. The energy used for maintenance is lost from the body as heat.
- Synthesising new body tissues or milk. The energy expended is lost as heat.
- Storage of energy in the tissues (fat, lean), or secretion as milk.
- Maintenance of body temperature in a cold environment.

A summary of estimated energy requirements of pigs is given in Table 6.5. These values can be used in a factorial approach to estimate the energy requirement for prescribed levels of production.

Protein/amino acids

Protein is required in pig feed to replace body protein that is continually being lost from the body (protein turnover) and to supply new protein for lean growth or milk production. The pig requires amino acids (AAs) rather than protein

Table 6.5 **Energy requirements of pigs**

Process	DE requirement
Maintenance	0.5 MJ/kg $W^{0.75}$
Growth	15–25 MJ/kg liveweight gain
Protein deposition	44 MJ/kg protein
Lipid deposition	50 MJ/kg lipids
Cold thermogenesis	0.017 MJ/kg $W^{0.75}$/°C below LCT
Maternal growth during pregnancy	21 MJ/kg liveweight gain
Uterus and fetal growth:	
70-day gestation	0.53 MJ/day
100-day gestation	2.2 MJ/day
Milk production	8.7 MJ/kg milk

DE = digestible energy; LCT = lower critical temperature; MJ = megajoules; W = weight.

Table 6.6 **Ideal amino acid balance for the growing pig**

Amino acid(s)	Liveweight				
	5–20 kg (Baker & Chung, 1991)	**20–50 kg** (Baker & Chung, 1991)	**20–100 kg** (ARC, 1981)	**20–100 kg** (Fuller et al., 1979)	**50–100 kg** (Baker & Chung, 1991)
Lysine	100	100	100	100	100
Methionine + cystine	60	62	50	53	65
Threonine	65	67	60	55	70
Tryptophan	18	19	15	12	20
Isoleucine	60	60	55	50	60
Leucine	100	100	100	99	100
Histidine	32	32	33	33	32
Phenylalanine + tyrosine	95	95	96	99	95
Valine	68	68	70	70	68
Non-essential AAs			750		

Sources: ARC = Agricultural Research Council.

per se. There are 10 AAs that are considered essential, meaning that they cannot be synthesised by the animal and must be supplied in the diet. Non-essential AAs can be synthesised from other AAs or simple nitrogenous precursors. The AAs must be present at the site of protein synthesis in the required proportions (amino acid balance). Amino acids cannot be stored for long periods of time. If an essential AA is limiting (supplied at below requirement level), then protein synthesis will be impaired; i.e. the utilisation of all AAs will be impaired. The ideal AA balance for pigs is given in Table 6.6.

The ideal balance is regarded as being optimal regardless of sex, breed or strain. Rather than considering levels of AAs, reference is usually made to levels of ideal protein. Lysine is often the first limiting amino acid in pig diets, and thus it is common to refer to lysine levels as a proxy for balanced protein.

The AA requirement for different body functions and the AA post-absorptive efficiency of utilisation are presented in Table 6.7. When proteins are used as a source of energy by the pig, the net energy provided is 11.5 MJ per kilogram of protein.

Diets are usually formulated using ileal digestibility amino acid values; these can be either apparent or corrected for endogenous losses (standardised or true ileal amino acid digestibility).

Feed ingredients

The main sources of energy are cereals (barley, wheat, maize) and tallow. These compounds are deficient in AAs, so protein-rich ingredients are mixed with the cereals. The main sources of feedstuff available in New Zealand, and their limitations of use, are given in Table 6.8. There has been a recent increase in the use of by-products from the food industry — whey, dissolved air flotation (DAF) product from the dairy processing industry — as well as products unsuitable or past their 'best by' date for human

Table 6.7 **Amino acid requirement of different classes of pig, and coefficients of utilisation**

Amino acid(s)	Maintenance		Growth		Milk (g/kg milk [5% protein])
	Pregnancy[1] (mg/kg $W^{0.75}$)	Growing pig (mg/kg $W^{0.75}$)	Sow and conception products (g/kg weight gain)[2]	Growing pig (g/kg protein)	
Lysine	25	66	12.1	70	3.9
Methionine + cystine	26	33	5.08	35	1.7
Threonine	39	39	6.4	42	2.1
Tryptophan	5	9	1.5	10	0.6
Isoleucine	30	36	6.6	38	2.3
Leucine	20	66	11.3	70	4.3
Histidine	0	2	6.6	23	1.4
Phenylalanine + tyrosine	46	63	11.4	67	4.1
Valine	0	46	7.8	49	2.8
Coefficient of utilisation	1	1	0.635	0.85	0.8

g = grams; kg = kilogram; mg = milligrams; W = weight. 1. Based on Baker et al. (1966) and Baker & Allee (1970).
2. Based on Speer (1990).

consumption, such as bread, yoghurt, fruit and other grocery and bakery items. These allow the pig farmer to reduce their feed costs while also reducing the environmental impact of wasted human food products.

The use of swill (kitchen, restaurant or hospital food waste) is not recommended for pigs. By law in New Zealand, all food waste that contains or has come into contact with meat must be treated before it is fed to pigs. This means either heating it to 100°C for one hour, or treating it to a standard approved by MPI's Director-General and notified in the *New Zealand Gazette*. The treatment requirement applies to all food waste that contains raw or cooked meat or has come into contact with raw or cooked meat. It applies to both commercial and household food waste. The regulations define 'meat' as any material taken or derived from an animal, with the exception of egg, milk and rendered material (such as tallow, blood meal, meat and bone meal). Egg and egg products, milk and milk products, and rendered material do not need to be heat treated.

Pig diets are also supplemented with a range of feed additives, including the following.

- **Synthetic AAs** — lysine, methionine, tryptophan and threonine can be purchased in bulk and are added to correct AA deficiencies.
- **Vitamins** — appropriate mixes are included in feeds in excess of requirements; extreme vitamin deficiencies are relatively rare, but could be induced by interaction with other dietary components (e.g. vitamin E/fat).
- **Minerals/trace elements** — always included, and are rarely limiting to production; the

Table 6.8 **Main ingredient sources for pig diets in New Zealand**

Feedstuff	Main nutrient	Reason for limit	Upper limit (% of diet)	
			Grower	Sow
Barley	Energy	–	–	–
Maize	Energy	Palatability, mycotoxin, high polyunsaturated fatty acids	25–30	25
Wheat	Energy	Low in fibre, price	–	–
Bran	Energy	High in fibre, bulky	5	15
Pollard	Energy/protein	High in fibre	10	15
Peas	Energy/protein	Anti-nutritional factor	15–20	10
Soybean meal	Protein	Price	–	–
Meat and bone meal	Protein	High ash content, mineral and amino acid imbalances	10–15	5–10
Blood meal	Protein	Palatability, amino acid imbalance	5	5
Fish meal	Protein	Fish taint, high polyunsaturated fatty acids, high mercury level in New Zealand fish meal	7	7
Milk powder	Protein	Price	–	–
Full-fat soybean meal	Protein/energy	High energy, high polyunsaturated fatty acids	10	10
Soybean oil	Energy	High polyunsaturated fatty acids, mixing problems	5	5
Tallow	Energy	Mixing problems	5	5

Source: Morel et al. (1999).

balance between the minerals needs to be checked.

- **Antioxidants** — these should be included in high fat/oil diets.
- **Flavours** — many flavour compounds have been used, and are advocated, particularly in feeds for young pigs to enhance appetite.
- **Enzymes** — a wide array of enzyme compounds are available commercially, to increase the digestibility of the complex polysaccharides in the diet (although this is not always successful).
- **Probiotics and prebiotics** — these products may alter the pigs' gut microflora, change the pH in their gut and enhance health; although this is not always successful, the importance of these additives will increase in the quest to reduce the amount of antibiotics in pig diets.
- **Copper** — this element has long been known to have a growth-promoting effect; it is routinely added to diets at levels higher than the pig's requirement for growth (125–250 ppm), but is not included in breeder diets as it accumulates in the liver over time.
- **Zinc** — included in piglets' diets at high dietary levels (2000–3000 ppm) to reduce the incidence of diarrhoea and increase weight gain in newly weaned pigs

- **Growth promoter** — repartitioning agents have been shown to increase growth rate and carcass leanness; these products are approved in New Zealand for commercial use (e.g. porcine growth hormone, ractopamine [Paylean]), although NZPork does not recommend their usage.
- **Antibiotics/medicaments** — an array of these compounds has historically been added to feeds, sometimes to combat disease and sometimes to enhance growth; nowadays, they can only be used for their therapeutic effect.

Diet formulation and mixing

Special diets are formulated for the creep feeding of young piglets (feeding piglets a special diet which the sow cannot reach), and for weaners, growers, finisher pigs, gilts, dry sows, lactating sows and boars.

Diets are usually complex mixtures of several ingredients, and when fed at the prescribed levels should meet the nutrient requirements for each class of pig. There is considerable variation in the recommended allowances for specific nutrients published in the literature, as pig requirements depend on factors such as age, sex, breed, environment and feeding level. In 2012, the National Research Council (NRC, 2012) published a revised edition of the book *Nutrient requirements of swine*; a nutrient partitioning modelling approach is used to determine nutrient requirements for growing–finishing pigs, gestating sows and lactating sows. The various models are available in a spreadsheet that can be downloaded from http://dels.nas.edu/Report/Nutrient-Requirements-Swine-Eleventh-Revised/13298.

Many diets made of different feedstuff mixtures will allow the pigs' requirements to be met, but there is a unique diet that will do so at the lowest cost. The linear programming technique is used to find the least-expensive diet that meets the nutrient requirement for each class of pig.

The processing of a diet takes place via a series of steps:

- Cereals are finely ground (hammer-milled) to enhance digestibility, although if the grind is too fine then animals become predisposed to gastric ulcers.
- Protein sources are added.
- Additives are bulked up, then included.
- The diet is mixed well.
- Sometimes the mixed diet will be pelleted (compressed into pellets at high pressure/temperature). However, this carries with it the possibility of damage to AAs and vitamins.

Feeding level

Sow productivity (number of pigs per sow per year) is economically important, and is directly under the control of the farmer through nutrition, management, health and housing. Over the past decade, the genetic make-up of sows (and boars) has changed considerably. Sows are leaner, have larger litters, grow faster, have a reduced appetite and produce more milk. A body fat content in the sow of lower than 10 per cent is associated with reproductive problems. Feeding the modern, leaner sows according to liveweight will result in a percentage body fat that is below the critical 10 per cent after a few farrowings. Such sows will not come back on heat and are likely to be culled. This phenomenon is known as thin sow syndrome. It is important to feed sows according to their fat reserves, which are

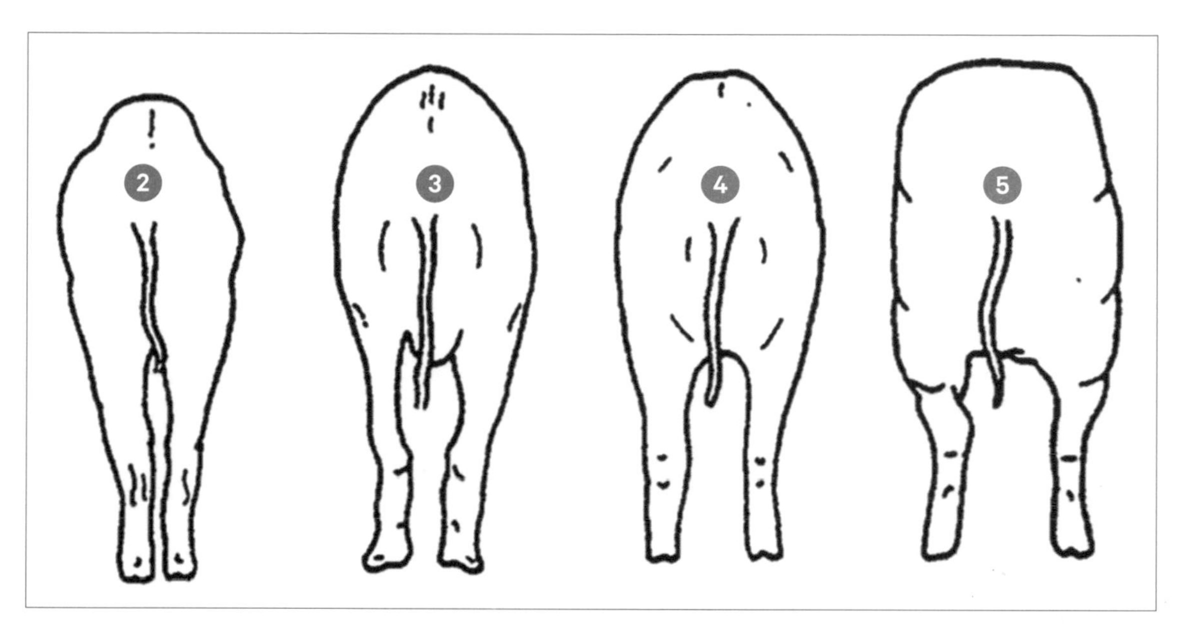

Figure 6.8 Visual condition scoring scheme using a diagrammatic standard.

usually estimated by using a condition score (Figure 6.8).

Feeding recommendation for gilts

The attainment of sexual maturity (puberty) is delayed when feed intake is very low (e.g. < 50 per cent of ad libitum intake). There is some evidence that a high plane of nutrition from selection of replacement gilts through to mating may lead to an increased incidence of silent heats. However, gilts fed ad libitum produce more ova. There is evidence of a flushing effect on ovulation (a high feed intake 10–14 days before oestrus promotes ovulation). It is thus common practice not to mate gilts on their first heat period but during their second or third heat period, to take advantage of the flushing effect and achieve a larger litter size (an additional piglet).

High intakes of energy after mating may reduce implantation of embryo in the uterus, increase embryo mortality and increase the loss of body weight during lactation.

From selection (90 kg) to second oestrus, gilts are fed 3 kg/day of a high-quality diet depending on leanness, then restricted to 2.3 kg/day for 10 days and flushed with 3–3.5 kg/day in the period leading up to mating. The gilts are mated at the third oestrus (125 kg, 220 days of age), and then a restricted feeding pattern (2.2–2.5 kg/day) is adopted for three weeks after mating to ensure high rates of implantation and embryo survival. With modern gilt genotypes, however, this may not be necessary. A reduced feed intake before mating not only reduces ovulation rates but some gilts may also not show oestrus.

From 21 days after mating to three weeks before farrowing, gilts are fed to reach a condition score of 3.5 to 4 (2–2.5 kg/day). Three weeks pre-farrowing, an increase in feed intake will increase litter weight and decrease piglet mortality. However, the risk of farrowing fever (MMA) is increased. Feed is restricted slightly (around 3–3.5 kg/day) for one week following farrowing, but from one week pre-weaning sows are fed ad libitum (6–8 kg/day).

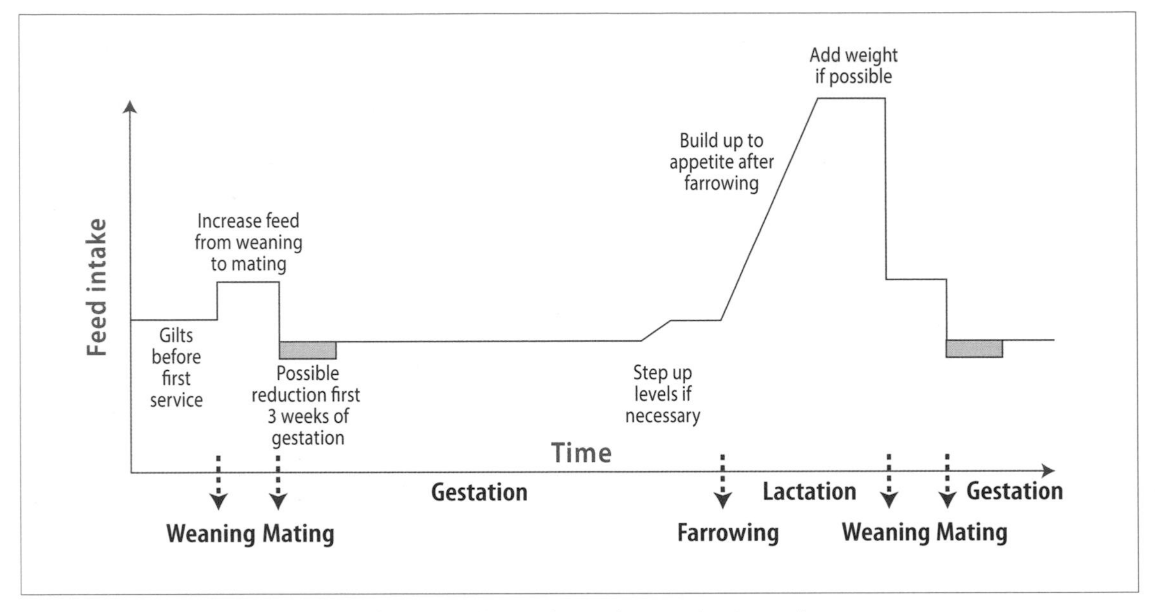

Figure 6.9 Changes in feed allowance for gilts and sows during the reproductive cycle.

Feeding recommendation for sows

Sows are fed to meet a body condition target, rather than being fed a prescribed amount of meal. Typical target scores (on a 1–5 scale) are 3 at weaning, 3.5 at mid-pregnancy and 3.5–4 at farrowing. The aim is to have a herd average score of 3.5. Feeding levels are increased for scores below 3.5 and decreased for scores above 4. The degree of change in feeding will depend on several factors, such as environmental temperature, nutrient density of feed, and health status. Excessively high feeding levels during pregnancy may reduce embryo survival and decrease lactation appetite. Sows gaining a lot of liveweight or a lot of fat during pregnancy also tend to lose more during lactation.

The approximate feeding levels with a barley-based diet are:

- **Weaning to mating:** for a prompt oestrus and a high ovulation rate, 3.5–4 kg/day in one meal is fed until after mating. The lighter, thinner sows will adjust their state by increasing the weaning to conception interval.
- **Mating to three weeks post-mating:** 2.3–2.5 kg/day.
- **Three weeks post-mating to three weeks pre-farrowing:** 2.3–2.5 kg; feeding is according to condition, with a target score of 3.5–4 at farrowing being the aim. Feed levels are increased for sows in poor condition.
- **Last three weeks of gestation:** an increase in feed intake to 3–3.5 kg/day will increase litter weight and decrease piglet mortality; to reduce the risk of farrowing fever, feed intake is then reduced to 2 kg/day for the five days before farrowing. High-bran feeds can be provided for laxative purposes.
- **Lactation:** in the first week around 3 kg/day is fed, and then sows are fed to appetite (6–8 kg/day) in two feeds each day. Fats and oils can be added for lean sows or sows with large litters. Thinner sows will adjust their milk yield downwards, thus reducing

the piglets' growth rate. Sows that are in good condition and have not lost excessive amounts of fat during lactation will rebreed more readily than very thin sows.

A summary of the different feeding levels for gilts and sows is presented in Figure 6.9.

Feeding recommendation for piglets

Piglets have a high growth potential (they can double or triple their weight within 14 days of birth), and are adapted to the high-quality sow milk. Sow milk, with 20 per cent dry matter and 5.1 MJ/kg, is an energy-rich milk. Piglets start life with only 1–2 per cent of fat in their bodies; by three weeks of age, at least 16 per cent of the body weight of milk-fed piglets is usually made up of lipids (fats).

The milk yield of the sow is a function of the number of piglets suckling and the stage of lactation; peak lactation is usually at three weeks. Under current management practices piglets are weaned at 3–4 weeks, which corresponds to the optimal lactation length for maximising the number of piglets weaned per sow per year. Natural weaning occurs at 12 weeks.

The piglet's gut has high lactase activity but low amylase, maltase and sucrase activity. Thus, piglets are only capable of using lactose and glucose in early life; supplementary feeding should take this into account. As the piglet grows, the milk yield falls below the growth potential at about three weeks of age. Supplementary feeding (creep feed) of suckling piglets is necessary for:

- litters of reasonable size, to obtain the greatest possible growth
- all litters, to accustom piglets to the nature of solid food in order to prepare for early abrupt weaning

Nutrients must be in concentrate form, highly digestible and palatable, and preferably contain milk proteins. Creep feed is usually provided from the second week of age.

Feeding recommendation for growing pigs

Feeding strategies for growing pigs can be defined as a combination of:

- feed form — mash diet, pelleted diet or liquid feed
- feeding intensity — from restricted feeding to ad libitum feeding
- number of diets fed from weaning to slaughter (as the amino acid and energy requirements change as the pig is growing) — one diet only, phase feeding (weaner, grower, finisher), or blend feeding (mixing a high-nutrient diet and a low-nutrient diet to match pig requirements on a weekly basis or more often)
- amino acids and energy concentration of the diets — different genotypes will have different requirements at different stage of growth

Clearly, there are many possible feeding strategies for growing pigs, but only one will result in a maximum profit (gross margin) per pig or per pig place per year. As both feed cost and price schedule change over time, the optimal feeding strategy will also change. However, as leaner pigs fetch a higher price per kilogram of carcass and also have a better feed conversion ratio, finding a strategy that maximises lean growth rate generally maximises profitability. Thus, being able to quantify the lean growth potential of commercially used pig genotypes is required when establishing their nutrient requirements and designing feeding strategies that will maximise farm profitability.

The main drivers of lean growth potential are the maximum protein deposition potential (PdMax) and the minimum lipid to protein ratio in the whole body (Target L/P). PdMax represents the maximum daily protein deposition rate (g/day) that can be achieved for a certain type of pig, and Target L/P represents the energy partitioning between lipid and protein deposition when energy intake, but not protein intake, is limiting. In 2012, the PdMax value for the main pig genotype used in New Zealand varied between 160 and 200 g/day for female pigs and between 180 and 240 g/day for entire male pigs.

Using this type of information, pig growth models like those in the NRC spreadsheets mentioned previously could be used to determine the economically optimal nutrient requirements for the growing pig.

Pork quality

Meat quality is determined by breeding and husbandry practices, in particular by the handling of animals immediately before and carcasses immediately after slaughter. There is some concern that the eating and manufacturing quality of pork may be declining. Organoleptic problems including dryness, hardness, inferior flavour and changes in colour, texture and water-binding capacity have been observed. The eating quality of pork is an individual experience that depends on its appearance and palatability. Fat contributes to the eating quality of pork, as it carries much of the typical flavour found in meat.

There are seven important traits that affect eating quality: meat colour, amount of intramuscular fat, water-holding capacity, tenderness, pH, polyunsaturated fatty acids, and skatole and androstenone level. These characteristics influence the decision of the consumer to buy the meat. They can be assessed subjectively by a taste panel made up of trained or untrained people, or by objective measurements of related physio-chemical properties. A review of pork meat quality in New Zealand has recently been published (Morel et al., 2010).

- **Meat colour:** meat that is too pale or too dark (classified as pale soft and exudative [PSE] and dark firm dry [DFD], respectively) is not attractive to the consumer. The amount of muscle pigment and the changes in pH level after slaughter will influence the colour of the meat.
- **Intramuscular fat (IMF):** this is the fat located inside the muscle. The percentage of IMF in a muscle will influence meat appearance (marbling) and also have an effect on meat tenderness, juiciness and flavour. An optimum level of 2.5 per cent IMF has been proposed.
- **Water-holding capacity (WHC):** this is the ability of meat to retain water under standard conditions. A low water-holding capacity is associated with a drier meat of

poorer appearance and a higher weight loss during processing.

- **Tenderness:** meat tenderness can be defined as the force needed to bite through the meat. An increased content of intramuscular fat and a higher water-holding capacity results in a more tender meat.
- **pH:** this is a measure of meat acidity. Meat colour and water-holding capacity are influenced by the pH changes in the muscle after slaughter. The pH can be measured just after slaughter (pH 45 minutes) or one day after slaughter (pH ultimate). Changes in pH after slaughter are discussed below.
- **Polyunsaturated fatty acids:** the fat quality of pork depends on the chemical composition of fat tissues and in particular the content of polyunsaturated fatty acids (e.g. linoleic acid C18:2). Increased polyunsaturated fatty acids in fat tissues results in a more oily consistency, a grey-yellowish colour and an increased tendency to produce rancid flavours with storage.
- **Skatole and androstenone:** these compounds are responsible for causing the off-flavour in pork commonly referred to as boar taint (discussed below).

Major pork quality problems

After slaughter, the transition of living muscle to dead meat is controlled by the glycolytic system, which converts glycogen to lactate, and by other processes that lead to dephosphorylation of ATP. Acidification (decreased pH) occurs as a consequence of the glycolysis, and rigor as a result of the loss of ATP. Rigor mortis takes place in pig muscle over a period of about 10 hours, during which the pH falls from about 7 to 5.5.

Occasionally this change may occur within a few minutes after death, and the meat will develop the PSE (pale soft and exudative) characteristic. When the pH falls below 6 and the temperature is above 30°C, an extensive denaturation of the soluble and structural proteins occurs that leads to a loss of their water-holding abilities. The precipitated protein interferes with the optical properties of the muscle, giving a paler appearance. In New Zealand the incidence of PSE has been reduced from 40 per cent in the early 1990s to 5–6 per cent today.

The dark firm dry (DFD) condition occurs when there is not enough glycogen available in the muscle at death to allow sufficient acidification to occur; the meat remains at pH 6–6.3. A high ultimate pH (24 hours after slaughter) will have a negative effect on the shelf life of the pork, as a low pH level has an inhibitory effect on bacterial growth. In New Zealand the proportion of DFD pork is below 7 per cent.

Hampshire-type or acid meat has a poor water-binding capacity. This is associated with a very low ultimate pH, rather than rapid pH decline and protein denaturation as in PSE.

PSE, DFD and Hampshire-type (acid) meats can be detected through:

- measurements of meat pH at 45 minutes and 24 hours after slaughter — this is presently the most reliable method of detecting PSE meat
- measurement of electric conductivity in meat — this is not accurate 30–40 minutes after slaughter, but can be used from two hours up to 36 hours after slaughter to detect PSE meat
- measurement of light reflectance or scatter in muscle, using a carcass classification instrument like FOM or HGP — this is not

accurate 30–40 minutes after slaughter; however, measurement of colour with GOFO or Unigalvo instruments 24 hours after slaughter allows the identification of PSE and DFD meat

Genetic influences on meat quality

Both PSE meat and Hampshire-type meat are under the control of major genes.

The halothane gene is a recessive gene associated with malignant hyperthermia (MH); it results in PSE meat and the porcine stress syndrome (PSS). A single point mutation in the gene for skeletal muscle ryanodine receptor (RyR1) on chromosome 6 (substitution of a T for a C at nucleotide 1843) in pigs is correlated with malignant hyperthermia (MH).

The percentage of halothane-positive (homozygous-recessive) animals (those carrying two copies of the recessive gene) differs between breeds and lines. In the Landrace breed the incidence of halothane-positive animals may vary between 1 per cent and 90 per cent, and in the Piétrain breed between 80 per cent and 90 per cent. The frequency of halothane reactors is generally low in Large Whites (0–5 per cent) and Durocs (0–1 per cent). Halothane-positive animals have a 2–4 per cent higher lean content than halothane-negative animals; thus, selection on the basis of percentage lean will lead to an increase in PSE meat if meat quality is not also included as a trait in the selection process.

The Napole gene is associated with Hampshire-type meat; it is a dominant gene.

Environmental influences on meat quality

In general, chronic stress prior to slaughter leads to DFD problems and acute stress leads to PSE problems. The interaction between different handling procedures and genotype will affect how an individual pig will respond to stress. Depletion of glycogen by chronic stress will prevent the formation of PSE meat in animals that are potentially susceptible to PSE, but pigs that are very susceptible to stress may not respond favourably to ideal pre-slaughter handling because the trauma of the slaughter process itself is sufficient to initiate development of the PSE condition.

- **Transport:** the conditions of transport are more important than the journey time. Conditions such as ventilation, optimal stocking density (0.4–0.5 m^2/pig), careful driving and flat unloading should be implemented.
- **Lairage (resting after transportation and before slaughter):** longer lairage in a slaughter plant reduces PSE but increases DFD. The optimum lairage time of 4–6 hours reduces PSE without increasing DFD.
- **Stunning:** by CO^2 or electric. The physical spasms that occur during and following stunning can lead to broken bones, blood splash, bruising and PSE meat. Excessive stimulation by electric current can lead to the PSE condition, but it is unlikely that the high-voltage systems used today produce any more or any less PSE meat than the traditional low-voltage methods.
- **Chilling:** the rapid chilling of fast-glycolysing (PSE) muscles shortens the time during which low pH coincides with high temperatures, and therefore reduces drip loss and leads to meat with less pale colour. However, the quality attributes are still inferior to those of normally glycolysing muscles.

Boar taint

Rearing entire males instead of castrates is economically advantageous to pork producers: entire males grow faster, use their feed more efficiently and produce leaner carcasses. Boar taint is a distinct unpleasant (urine-like) odour that is sometimes present when fat or meat from boars is cooked. Not all pork from entire males will exhibit boar taint, and individual consumers may differ in their ability to detect taint. Nevertheless, boar taint is considered to be a major risk to consumer confidence in pork. The two compounds mainly responsible for causing taint are skatole, produced in the intestine by bacterial fermentation of the amino acid tryptophan, and 5-α-androstenone, a C19-steroid synthesised in the testes by the sexually mature boar.

In New Zealand, male pigs are slaughtered as entire. The industry trend towards producing heavier carcasses brings with it an increased risk of producing tainted carcasses. A survey conducted in New Zealand in 1996 showed that 13 per cent of fat samples exceeded the European Community (EC) threshold level for skatole (< 0.2 ppm), 31 per cent the EC threshold for 5-α-androstenone (< 1 ppm) and 19 per cent both EC threshold levels. This means that 63 per cent of the samples would have been considered having boar taint according to the EC standards of the time.

A similar survey was conducted in 2009: 75 per cent of fat samples from entire males had androstenone values below 0.5 ppm, 79 per cent had skatole values below 0.2 ppm, and 90 per cent were below both values. In total, 35 per cent of entire male pigs would be considered to have boar taint. This indicates that boar taint is still a major pork-quality issue in New Zealand, limiting the domestic consumption of pork, especially by new migrants. A common remark by migrants is that they have tried New Zealand pork but don't like its taste, and therefore don't buy it anymore.

There are different ways to manage boar taint. Surgical castration is used worldwide and is allowed in New Zealand for piglets aged less than seven days. However, in the EU, due to animal welfare concerns surgical castration should be done under anaesthesia. An alternative is the use of immuno-castration; the vaccine Improvac is available for use in New Zealand. Keeping pigs clean is another way to lower the skatole levels, as well as different dietary manipulations (adding inulin in the diet or reducing the amount of dietary tryptophan). In female pigs, high skatole levels will also cause an off-flavour.

Intramuscular fat

Intramuscular fat is located inside the muscle; it influences meat appearance (marbling) and has an effect on meat tenderness, juiciness and flavour. A minimum level of 2 per cent IMF in the longissimus muscle has been proposed. The IMF percentages (IMF%) reported for New Zealand pork were over this threshold in the 1990s, but values reported recently for modern pig genotypes are between 1 and 1.5 per cent. This reduction in IMF% over time could be associated with the genetic selection for high carcass leanness or low carcass backfat thickness. The lower than optimal level of IMF% in the loin in New Zealand pork could limit the acceptability of sensory characteristics that can be achieved. Entire males have less IMF than female pigs, which have less than castrated male pigs. Duroc pigs have a higher content of IMF than Large White and Landrace or Piétrain pigs.

Free-range pigs.

IMF has a medium to high heritability that can be improved genetically.

Fatty acid profile

There is increasing consumer awareness of improved health and wellbeing through good nutrition. The potential for marketing health improvement through dietary ingredients has been recognised by the food industry. Consumers are demanding pork with less fat, and at the same time they are very aware of the type of fat certain foods contain.

Omega-3 (n-3) polyunsaturated fatty acids (PUFAs) have a wide range of beneficial effects on human health, especially against heart disease. In addition to the concentration of some specific fatty acids in fat, their relative and comparative proportions are also relevant for human health. The ideal ratio of PUFAs to saturated fatty acids (SFAs) for human nutrition is 0.7, and the ratio between C18:2 (linoleic) and C18:3(n-3) (linolenic) should be below 6.0.

Modifying the fatty acid profile of muscle and adipose tissue by dietary means is easier in monogastric species than in ruminant species. The influence of nutrition on fatty acid composition in monogastric species has been widely documented since the 1930s. In pigs, the fatty acid profile varies between tissues, breeds and gender. In New Zealand, pigs are traditionally fed on cereals supplemented with protein and fat from either animal or plant sources. These ingredients contain many and differing levels of fats and proportions of fatty acids. Cereals and other plant ingredients tend to contain more PUFAs, while ingredients from animal sources tend to contain more saturated fatty acids.

Feeding fish oil, which is rich in 'healthy' fatty acids (eicosapentaenoic acid, C20:5(n-3) and docosahexaenoic acid, C22:6(n-3)), to pigs increases the concentration of these fatty acids in pork. However, the inclusion of fish oil in pig diets may cause fishy or rancid flavours and soft fat, and it is expensive. Similar 'health' effects can be obtained by including other PUFA sources such as linseed oil, soybean oil, palm oil or canola oil in pig diets. Such inclusion will increase the content of α-linolenic acid, which is a precursor of C20:5(n-3) and C22:6(n-3).

There is some conflicting information on how dietary fatty acids are incorporated or eliminated from pork tissue. Some studies have shown that the final level of PUFA in pork tissue depends on how long the diet has been given (accumulation), and when it was fed last (elimination), while in other studies no elimination process was demonstrated. Overall, it seems that the fatty acid profile in pork fat at slaughter is primarily influenced by the total dietary PUFA intake and less by the time at which high-PUFA diets are fed.

A potential problem for pork and pork products with a high PUFA content is that they have lower oxidation stability. Lipid oxidation is one of the major causes of off-odours and off-flavours in pork. It can be assessed by measuring malondialdehyde compounds that react with thiobarbituric acid (TBARS number). It is unlikely that consumers will detect off-flavours when the TBARS value is below 0.5 mg MDA/kg. The oxidation stability of pork products can be increased by the addition of antioxidants such as vitamin E (α-tocopherol) in the diet. Overall, increasing the PUFA content, especially *n*-3 fatty acids, in pork and pork products has a positive effect on their nutritive value and a negative effect on their processing and shelf-life properties.

Genetic selection for meat quality

Over the past decades, an impressive decrease in backfat thickness has been observed in pigs in New Zealand. This selection for leanness is likely to be accompanied by a decline in eating quality. A summary of genetic parameters for eating quality traits is presented in Table 6.9.

Quality traits show medium (WHC) to high (IMF and C18:2) heritability values. On average, all quality traits are unfavourably genetically related to backfat thickness and percentage lean. Average daily gain is unfavourably genetically correlated to colour and ultimate pH and favourably correlated to IMF and C18:2; IMF is favourably related to tenderness and fat quality. This means that selection on low backfat thickness (high lean content) and high growth rate, without consideration of any meat quality traits, is more than likely to cause a decline in eating quality.

Table 6.9 **Heritability values (*h2*) for meat colour (Colour), intramuscular fat (IMF), water-holding capacity (WHC), tenderness (Tend), pH ultimate (pH) and linoleic acid (C18:2) and their genetic correlations (*rg*) with average daily gain (ADG), backfat thickness (BF), lean percentage (Lean) and intramuscular fat (IMF)**

	Colour	IMF	WHC	Tend	pH	C18:2
h2	0.30	0.50	0.20	0.30	0.20	0.55
Range	0.10/0.60	0.25/0.85	0.05/0.65	0.20/0.40	0.10/0.40	0.40/0.70
rg ADG	−0.15	0.40	0.00		−0.15	−0.20
Range	−0.50/0.15	0.15/0.60	−0.80/0.50		−0.40/0.50	−0.40/0.0
rg BF	0.20	0.20	0.20	0.20	0.15	−0.70
Range	−0.20/0.60	−0.20/0.60	−0.20/0.80	−0.20/0.40	−0.40/0.50	−0.50/−0.90
rg Lean	−0.25	−0.20	−0.35	−0.20	−0.10	0.60
Range	−0.60/0.20	−0.50/0.40	−0.80/0.25	−0.60/0.20	−0.70/0.35	0.40/0.80
rg IMF	−0.20		−0.05	0.25	0.0	−0.30
Range	−0.10/0.35			0.20/0.30	−0.20/0.4	

Source: Morel (1995).

Animal welfare

The intensive indoor farming of any animal reduces its ability to engage in a wide range of behaviours which would be possible outdoors. However, it may allow easier inspection, closer management, protection from excessive cold and heat and predation, and better health for the animals involved. Intensive farming of pigs has led to much attention by animal rights organisations.

The significant welfare issues for piglets include teeth clipping, tail docking, early weaning and high perinatal mortality. However, male pigs are not usually castrated in New Zealand. Teeth clipping is done to reduce damage to other piglets and sows. Growing pigs may engage in tail-biting, which can cause an increase in aggression and spinal abscesses in the victims; tail docking prevents tail-biting. The perinatal mortality of piglets is reduced if sows are held in farrowing crates, although these may be banned in the future due to the limiting of sow motility. Transport to slaughter may be a stressful experience for pigs, but appropriate stocking density in trucks and careful driving reduces this.

Aggression between sows is a problem when they are grouped after mating. Sow aggression was eliminated with the use of sow stalls, but as these are now banned aggression between sows when they are mixed after mating requires careful management. Sows farmed outdoors may have nose rings placed to reduce pasture damage.

In New Zealand, animal welfare standards are controlled through the Animal Welfare Act 1999. The focus of this Act is to prevent cruelty to animals, making the animal owner completely responsible for their welfare. Penalties under the Act can include fines, imprisonment and confiscation of animals. The National Animal Welfare Advisory Committee (NAWAC) advises the Ministry for Primary Industries on welfare issues. The Animal Welfare (Pigs) Code of Welfare 2010 (NAWAC, 2010) details how pigs may be farmed. The most critical feature of the code is the reduction in the time a sow may spend in a dry sow crate and a farrowing crate. Sow stalls were banned in 2015 and producers have had to develop group housing systems for sows. This has necessitated new housing structures and husbandry skills.

The code of welfare for pigs provides a set of minimum standards covering feed, new-born piglets, watering systems, indoors-farming buildings and maintenance, indoors-farming space, indoors-farming temperature and air quality, outdoors-farming environment, farrowing, dry sow stalls, tethering, boars, elective husbandry procedures, restraint and handling and movement, weaning, health, inspections, pre-transport selection and stockmanship.

Piglets roaming outdoors.

Disease

On a pig farm, a large number of animals are usually kept together in the same area; thus, diseases can rapidly spread from one pig to the next. It is therefore important to maintain a high health status. Many diseases affect pigs; those most relevant to the New Zealand pork industry are outlined here.

The pig population in New Zealand is free from many contagious diseases. Lameness in sows is a problem, and there are some important infections that cause gastrointestinal disease in piglets, and respiratory, skin and reproduction problems.

Diarrhoea

In young pigs, from soon after birth to weaning, diarrhoea may be caused by bacteria (especially *Escherichia coli*), rotaviruses and parasites (coccidia).

- E. coli diarrhoea is found worldwide, and there are many strains and types of E. coli in pigs. E. coli diarrhoea may be prevented by vaccinating pregnant sows twice before farrowing. Antibiotics, given orally or by injection, are used to treat affected piglets.
- *Clostridium perfringens* type C (CpC) bacteria also cause diarrhoea. An antiserum given orally or by injection within two hours of birth provides immunity to CpC. Antibiotics can also be used. Sows may be vaccinated twice with bacterin-toxoid during pregnancy to prevent the disease.
- Diarrhoea may also be due to porcine proliferative enteritis (PPE) caused by *Lawsonia intracellularis* bacteria. In this disease, the mucous membrane of part of the small intestine and/or the large intestine is thickened. Prevention is difficult. Outbreaks may be related to environmental stresses, including mixing together of stock.
- Uncomplicated diarrhoea due to rotaviruses has low mortality. There are no specific treatments, but the severity of an outbreak can be reduced by keeping piglets in a dry, warm environment and feeding them well.
- The coccidial parasite *Isospora suis* causes diarrhoea in young piglets (1–3 weeks of age); it is less common in weaned piglets. Farrowing crates must be disinfected between farrowings to ensure control of coccidiosis.

Other bacterial diseases

Salmonellosis in pigs is due to any one of several *Salmonella enterica* bacteria. All pigs are susceptible, but weaned or grower-finisher pigs particularly so. Vaccines combined with cleanliness and proper husbandry and pig flow will prevent the disease. Antibiotics have also been used to prevent it.

Pneumonia is caused by the bacterium *Mycoplasma hyopneumoniae*. Affected pigs cough, grow slowly and have lower feed efficiency. Herds supplying genetic stock should not have *M. hyopneumoniae* and must have testing and biosecurity protocols applied to ensure freedom from the disease.

Pleuropneumonia is caused by the bacterium *Actinobacillus pleuropneumoniae* (APP). This disease is highly contagious, causing coughing and resulting in high mortality. It can be seen in all pigs but is most common in those aged 6–20 weeks. Antibiotics may be used to treat both affected pigs and those in adjacent pens.

Leptospirosis is a contagious bacterial disease of pigs caused by several serovars. Leptospirosis in pigs can cause a range of clinical signs, including abortion and sickness in young pigs. It may be controlled by vaccination and/or antibiotics.

Erysipelas is an infectious disease of mostly growing or adult pigs, caused by the bacterium *Erysipelothrix rhusiopathiae.* It may cause acute or chronic disease. Erysipelas is identified by diamond-shaped skin lesions and by enlarged joints, lameness and endocarditis. It is treated with antibiotics.

Exudative epidermitis (greasy pig disease), seen in pigs up to eight weeks of age, is caused by *Staphylococcus hyicus.* Treatment is not very effective but injections of antibiotics may be useful.

Other viral diseases

Porcine parvovirus (PPV) causes reproductive failure in sows that have not previously been exposed to the virus. It results in mummified fetuses, delayed return to oestrus, small litters and abortion. The virus is widespread and almost all herds are infected without obvious signs. PPV may be prevented by vaccination before breeding.

Porcine circovirus type 2 (PCV2) is associated with post-weaning multi-systemic wasting syndrome (PMWS), porcine dermatitis and nephropathy syndrome (PDNS), porcine respiratory disease coplex (PRDC) and, occasionally, reproductive failure. There is no specific treatment for pigs with PCVAD but it may be controlled by vaccination.

Other parasitic diseases

Sarcoptic mange is a common skin disease caused by the mite *Sarcoptes scabiei* var. *suis.* It occurs as itching and spots in all age groups, and is the most important external parasitic disease of pigs. Pregnant sows are treated with external sprays or injections of avermectin before entering the farrowing sheds. Eradication is possible in most pig houses.

Prevention of disease

Maintaining the health of pigs on a commercial farm depends on strict biosecurity (including limiting the access of people to the premises), a vaccination programme suited to the important diseases, and rapid diagnosis and treatment of sick animals. Regular health checks are necessary, as is a veterinary health programme to maximise health and minimise disease issues.

Weaner piglets on straw.

References

Agricultural Research Council (ARC). (1981). The nutrient requirements of pigs. Slough, England: ARC, Commonwealth Agricultural Bureau.

Baker, D.H., Becker, D.E., Norton, H.W., Jensen, A.H., & Harmon, B.G. (1966). Some qualitative amino acid needs of adult swine for maintenance. *Journal of Nutrition*, 88, 382–90.

Baker, D.H., & Allee, G.L. (1970). Effect of dietary carbohydrate on assessment of the leucine need for maintenance of adult swine. *Journal of Nutrition*, 100, 277–80.

Baker, D.H., & Chung, T.K. (1991). *Ideal protein for swine and poultry*. Biokyowa Technical Review no. 4 (p. 16). Chesterfield, Missouri: Biokyowa, Inc.

Blum, J. K., (1983). *Populationsanalyse der schweizerischen Schweinerassen*. Dissertation 7412. Zurich: Eidgenössischen Technischen Hochschule Zürich.

Chidgey, K.L., Morel, P.C.H., Stafford, K.J., & Barugh, I.W. (2015). A comparison of sow and piglet productivity and sow reproductive performance in farrowing pens and farrowing crates on a commercial New Zealand pig farm. *Livestock Science*, 173, 87–94.

Davidson, A.A. (2005). The use of baseline production data in order to benchmark New Zealand with key international trading partners. In X.J. Paterson (Ed.), *Manipulating Pig Production* (p. 167). Werribee, Victoria: Australasian Pig Science Association.

Fuller, M.F., Livingstone, R.M., Baird, B.A., & Atkinson, T. (1979). The optimal amino acid supplementation of barley for the growing pig. 1. Response of nitrogen metabolism to progressive supplementation. *British Journal of Nutrition*, 41, 321–31.

Holmes, C.W. (1982). Climatic environment and pig production. In *Improved pig performance by better feeding: papers from the 1982 Pig Production Course held at Massey University in August 1982*. Palmerston North: Department of Animal Science, Massey University.

Krieter, J. (1986). *Entwicklung von Selectionsmethoden für das Wachstums- und Futteraufnahmevermögen beim Schwein*. Heft 31 (PhD thesis). Kiel, Germany: Christian-Albrechts-Universität zu Kiel.

Morel, P.C.H. (1995). Meat Quality Series: Relationship between eating quality, backfat thickness and intramuscular fat. *MRC News Brief* (Massey University, New Zealand), April.

Morel, P.C.H., Pluske, J., Pearson, G., & Moughan, P.J. (1999). A standard nutrient matrix for New Zealand feedstuffs. Palmerston North: Massey University.

Morel, P.C.H., Purchas, R.W., & Wilkinson, B.H.P. (2010). Review of pork-quality studies in New Zealand. *Proceedings of the New Zealand Society of Animal Production*, 70, 261–65.

Morel, P.C.H., Schwoerer, D., & Rebsamen, A. (1988). Prise en considération de la graisse intramusculaire dans l'indice de selection de Sempach. *Der Kleinviehzüchter*, 36, 1341–50.

National Animal Welfare Advisory Committee. (2010). Accessed at www.biosecurity.govt.nz/animal-welfare/codes/pigs/pigs.pdf.

National Research Council. (2012). *Nutrient requirements of swine*. Washington DC: The National Academies Press.

NZPork. (2014). Good practice guide for nutrient management in pork production. Accessed at www.nzpork.co.nz/images/custom/the_good_practice_guide_to_nutrient_management.pdf.

Schmidt, E. (1988). Analyse des Wachstums und seine Beziehungen zum Schlachtkörperwert beim Schwein. Heft 48 (PhD thesis). Kiel, Germany: Christian-Albrechts-Universität zu Kiel.

Speer, V.C. (1990). Partitioning nitrogen and amino acids for pregnancy and lactation in swine: a review. *Journal of Animal Science*, 68, 553–61.

Wäfler, P. (1982). Indexselektion in entgegengesetzter Richung — ein Selektionsexperiment mit Schweinen. (Dissertation 6979). Zurich: Eidgenössischen Technischen Hochschule Zürich.

Chapter 7

Poultry Production

Velmurugu Ravindran

Chapter 7

Poultry Production

Velmurugu Ravindran

Institute of Veterinary, Animal and Biomedical Sciences
Massey University, Palmerston North

Introduction

The term 'poultry' collectively refers to domesticated birds raised for their meat and eggs, and includes chickens, turkeys, ducks and geese, plus minor species such as guinea fowl, squabs, quails, pheasants and ostriches. Of these, the most prevalent worldwide are chickens and consequently the terms 'poultry' and 'chickens' are normally used interchangeably.

The commercial chicken sector falls into three distinct categories: meat production, table egg production and hatching egg production. The third sector, which includes breeder farms and hatcheries, is involved in the production of eggs for hatching, provides day-old chicks and links the meat chicken and layer industries. The first two sectors, which supply poultry products to consumers, differ widely in their structure, developmental history, production objectives and husbandry, and will therefore be addressed separately.

Consumers do not come into contact with the third sector and hence are unaware of it. The breeding sector imports stock, in the form of fertile hatching eggs, every 2–3 years. These eggs, which will form the great-grandparent generation, are initially held in quarantine farms under the supervision of the Ministry for Primary Industries (MPI).

The eggs are then hatched in company-owned quarantine farms under MPI supervision. These strict quarantine procedures protect the unique and superior health of the New Zealand poultry

Broiler chickens on deep litter. **Previous:** Two-week-old broiler chicks in experimental cages.

flock: New Zealand is the only country in the world that remains free of three major infectious diseases of poultry — infectious bursal disease, exotic Newcastle disease and the highly pathogenic avian influenza. This unrivalled low incidence of serious immune challenges is one reason for New Zealand achieving an internationally coveted poultry performance in terms of growth, feed conversion and egg production (Diprose, 2001).

The poultry industry has a domestic market focus, providing table eggs and meat to a population of 4.2 million. In recent years, however, export markets have begun to grow — chicken meat to Australia and speciality-type meat and eggs (free-range and organic) to Asian countries. Due to the superior health status of chickens in the industry, no imports into New Zealand of fresh or frozen poultry meat, or table eggs, are permitted. The major chicken-production areas are located around the larger population centres — Auckland, Waikato, Taranaki, Manawatu, Canterbury and Dunedin.

In New Zealand, the production of meat chicken and layers comes under the purview of the relevant welfare codes of 2012 (National Animal Welfare Advisory Council [NAWAC], 2012a, b). These codes set out minimum standards and recommended best practices, which are intended to encourage the industry to adopt a standard of husbandry, care and handling relating to all aspects of the care of meat chickens and layers. These minimum standards have a legal effect under the Animal Welfare Act 1999.

Glossary

Alternative systems: egg-farming systems that are not conventional cage systems, such as colony cages, aviaries, barns or free-range systems.

Aviary: a building for layer hens without conventional cages, similar to a barn but providing two or more floor levels and usually including deep litter, perches and slatted platforms.

Beak-trimming: removal of a portion of the upper beak to reduce feather pecking, cannibalism and egg eating; accomplished with a machine that cuts and cauterises the beak. It is often done during the first week of a chick's life.

Breeder farm: a farm that produces hatching eggs from breeder stock; it supplies eggs for hatcheries to produce day-old chicks that are used commercially for chicken meat or table egg production.

Broiler: a young meat-type chicken.

Broiler breeder: birds that produce hatching eggs for the supply of day-old chicks to commercial broiler production.

Brooding: the provision of warmth and the management of chicks from one day old to four weeks of age.

Cannibalism: a behaviour pattern that includes pecking of pen-mates, especially around the tail or vent, and resulting in various degrees of injury or in death. Often due to nutritional imbalances and management problems.

Conventional cage: a metal enclosure, usually made of wire, used to keep hens or growing pullets. These cages do not have a perch, a nest box or a dust-bathing area.

Colony cage: a modified and enlarged cage with more space than a conventional cage and with perching, nesting and dust-bathing areas. Also known as a furnished or enriched cage.

Controlled environment: an enclosed, insulated building containing pullets or layer hens, which provides total control of lighting, ventilation and temperature under automated control.

Crumbles: feed that has been pelleted and broken up (*see also* Pellets).

Day length: the period of time between the dark/light interface and the light/dark interface that the birds sense as the light period.

Designer eggs: eggs produced to contain higher levels of specific nutrients, such as selenium, omega-3 fatty acids, etc. Also known as enriched eggs.

Dual-purpose breed: a breed that has relatively good egg production as well as moderate fleshing. The males are usually grown out for meat purposes.

Egg sequence: number of eggs laid on consecutive days before skipping a day or more.

Fertility: the proportion of eggs in which fertilisation has occurred. Expressed as a percentage of eggs set in the incubator.

Floor eggs: eggs laid on the floor rather than in nest boxes of floor-type laying or breeder houses.

Forced moult: the moulting of feathers intentionally induced by changes in environment or diet. This physiological process is involved in the rejuvenation of the reproductive system to restore egg production rate and quality, thus extending the productive life of the hen.

Free-range: a farming type that provides birds with access to an extensive outdoor area which typically includes housing (either fixed or movable) similar to a barn, aviary or perchery without conventional cages.

Grower diet: the diet given to growing birds, usually following the starter diet (see below). There may be multiple grower diets that vary in formulation, based on differences in nutrient requirements due to age and other factors.

Hatchability: the proportion of eggs that hatch chicks. Expressed as a percentage of fertile eggs set in the hatchery.

Hen: an egg-producing bird from 18 weeks old to the end of laying.

Hock burns: irritation of the skin of the hock joint due to contact with irritants, usually wet litter.

Layer breeder: a bird that produces hatching eggs for the supply of day-old chicks to commercial table egg production.

Litter: wood shavings placed on the floor of the shed to absorb manure and moisture.

Mash: feed ingredients ground into small particles and mixed to provide a complete diet.

Meat and bone meal: a feed ingredient produced by cooking, drying and grinding animal products (including bones) not used for human consumption. Good source of protein, calcium and phosphorus.

Moulting: natural shedding of old feathers followed by their regrowth.

Organic production: production of poultry without the use of chemical inputs and with feeding of certified organic feed (feed ingredients and pasture are produced without synthetic fertilisers and pesticides).

Pellets: feed that has been steam-compressed into a small, rounded form.

Point of lay: the time at which a bird is physiologically ready to begin egg production.

Processing: refers to the slaughter, defeathering, evisceration, dressing and cutting into portions of the carcass. Further processing may proceed through cooking and value-adding.

Pullet: a young hen aged from eight weeks old to point of lay (see above).

Spent hens: laying hens that have completed their laying cycle.

Starter diet: diet formulated to meet the requirements of young chicks.

Withdrawal diet: the diet fed to birds for a specified time prior to slaughter, which contains no antibiotics or other compounds that are undesirable in chicken meat for human consumption.

A brief history of the poultry industry

The chicken was first domesticated approximately 10,000 years ago; ever since then, eggs and chicken meat have been an important part of the human diet. The time of introduction of poultry into New Zealand is a matter of conjecture, but it is very likely that missionaries in the Bay of Islands, arriving in 1814, were the first recorded poultry farmers (Hall & Clarke, 2011). Some reports do suggest that poultry were brought to the country in 1773, 1774 and 1777 by the explorer James Cook (Thomson, 1922), who gave hens to Māori in both the South and the North Island (Wintle & Lepper, 2012).

Early poultry production consisted of households having a few hens in the backyard to supply eggs for household consumption and some local retail egg sales. Increasing interest in having poultry was evident from the census data of estimated bird numbers between 1861 and 1871, during which time the numbers owned by European settlers increased from 236,098 to 872,174 (Binney et al., 2014).

At the turn of the twentieth century, the New Zealand government began to encourage domestic egg production because of the high nutritive value of eggs and the ease of setting up coops in backyards (Wintle & Lepper, 2012). Several state poultry stations were established to import, evaluate and breed improved stock. Two books were also published during this period, providing detailed information on improved production techniques that were available in the United Kingdom at the time, with the aim of promoting utility (dual-purpose) poultry farming (Ambler, 1923; Gordon, 1908). There is ample evidence from various sources dating back as far as the 1890s that many keen individuals in the poultry industry were applying scientific principles (Cundy, 2010).

In 1944, a Poultry Flock Improvement Plan was introduced by the New Zealand Poultry Board (NZPB), a statutory body established under the Poultry Runs Registration Act of 1933, and administered by the Department of Agriculture (Kissling, 1945; Wintle & Lepper, 2012). Eggs and chicks from these flocks were available to the public. The government aided the industry by providing field poultry instructors, a central demonstration poultry station and a disease diagnosis laboratory service. Poultry husbandry departments were also established in the two agricultural colleges (Massey and Lincoln; Kissling, 1945). As a result of these efforts, poultry farming gradually shifted from predominantly backyard production to small-scale production. The emphasis was on egg production; White Leghorns were the most popular breed, although heavy breeds such as the Orpington were recognised as being useful meat producers.

The production of eggs was not considered a commercial endeavour until the middle of the twentieth century, mainly due to the problems of disease outbreaks, poor genetics and the lack of understanding of improved technology. This situation changed after World War II, when increasing household incomes and the correspondingly greater affluence during the 1950s and 1960s resulted in increased demand for eggs and meat (Stone, 1972). Hens were moved indoors to barns with a deep litter, enabling better control of husbandry, and this resulted in marked increases in egg output, efficiency and profitability. Further improvements in economic efficiency were achieved during the 1960s, when hens were moved to cage housing to meet the increasing demand for eggs and to further improve the economic efficiency of egg production.

Initially, the structure of the commercial layer industry was regulated and controlled by the NZPB, which was charged with the development of an orderly marketing system. Controls were put in place on the number of birds a farmer could have, the number of eggs the farmer could produce, to whom eggs could be sold and the price that could be charged. The board's activities were funded by levies paid by owners of more than 25 layer hens (Kissling, 1945). In 1960, there were close to 4 million layers in the national flock, although half of these were still in backyard flocks of fewer than 25 birds. There were 1800 farms boasting over 100 birds. The cost of an egg was 4 shillings a dozen, equivalent to over $10 a dozen in today's money (Cundy, 2010).

Brown and white eggs have been proven to have the same nutritional value.

The years from 1953 to 1969 saw a steady increase in the number of birds farmed and eggs produced, with surpluses becoming a serious issue. During this period, farmers had applied scientific knowledge and improved the efficiency of egg production so dramatically that over-production continued to be a problem for some years. The end of the period also coincided with the lifting of restrictions to the importation of improved layer stock from the United Kingdom and Canada, in 1970. To curtail the extreme threat of surplus eggs, the NZPB introduced the Production Entitlement Scheme of 1970 and issued licences to keep hens. However, for a number of reasons this strategy was ineffective and consumers felt that prices were being kept high by the quota system. In 1985, a Government Commission recommended that the industry be exposed to the normal checks and balances associated with open competition; as a result, the NZPB was disbanded in 1989. This deregulation led to an immediate emphasis on the reduction of large-sized eggs and to increased volatility in egg returns and profitability. The table egg industry currently operates as a free market, with supply and demand determining prices. The outcome is that efficient producers are able to produce high-quality eggs at a reasonable price while inefficient producers are forced out of the business.

Production of chickens for their meat has traditionally been a sideline to the egg-laying business — often, a non-laying hen or a rooster was slaughtered. Chicken meat was a luxury, eaten

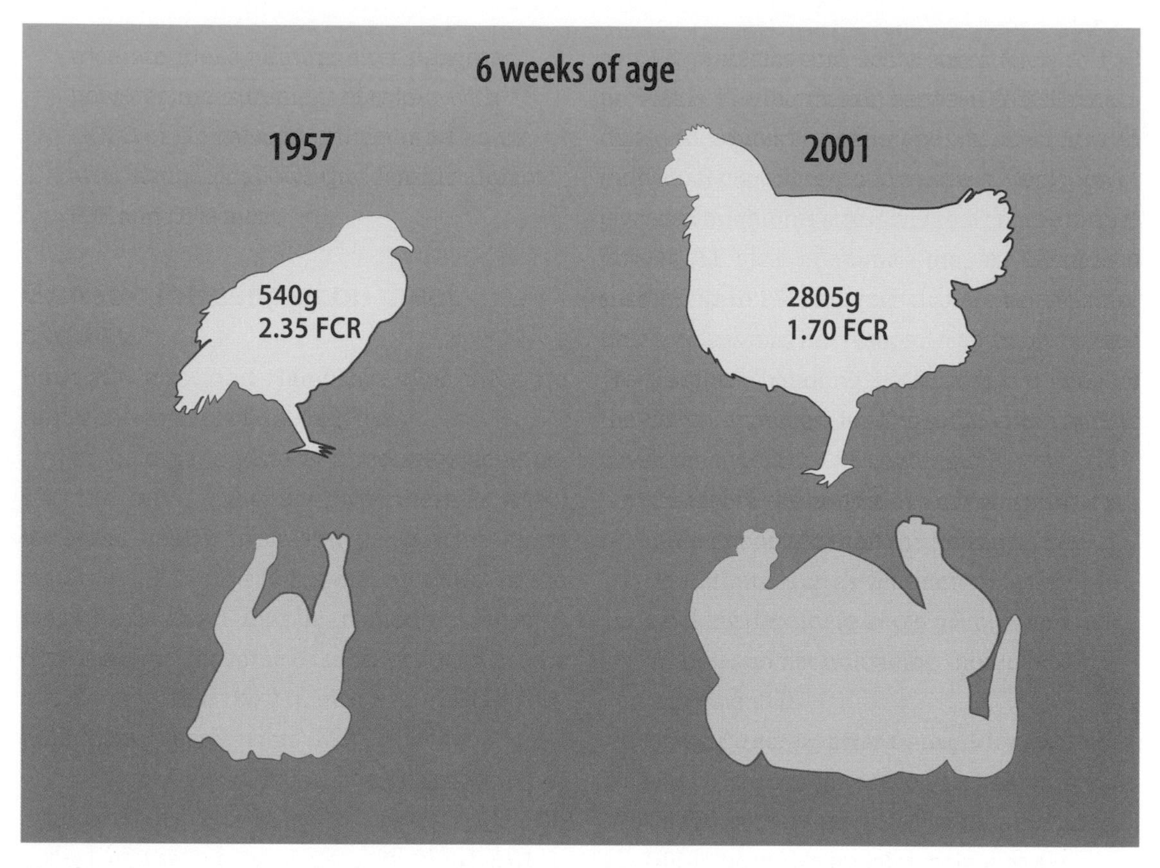

Change in the size of broiler chickens at six weeks of age over four decades. FCR = feed conversion ratio.

perhaps only once or twice a year on special days. During early times, owing to the plentiful supply of other meats, poor genetics and high production costs, the production of chicken meat was considered unprofitable (Kissling, 1945). The estimated per capita consumption in the late 1950s of red meats (beef and lamb) was 80 kg, pork 14 kg and poultry meat 1.8 kg.

Establishment of the commercial meat chicken (or broiler) industry began with the importation of Tegel breed grandparent stock in 1959, from Tegel Pty Ltd, an Australian chicken-breeding company. During the next three decades, broiler production underwent major changes in structure and technological advances, and evolved into a sophisticated industry.

Following the lifting of import restrictions from countries outside Australia in 1971, fast-growing broiler strains were introduced and vertical integration became the norm, with a single company involved in every stage of production, processing and marketing. Another significant development was the establishment of Kentucky Fried Chicken in 1971, and likewise other fast-food outlets, which had a major impact on the demand for and consumption of chicken meat. Today, about half of the chicken meat produced in the country is consumed through these outlets. While in 1959 no specialised broiler production existed, by 1989 some 42 million broilers were produced and marketed in a variety of forms — frozen, fresh and cut up. In 2015, 110 million broilers were processed in New Zealand (Brooks, 2016).

Chicken meat production

New Zealand has a small but highly organised and dynamic broiler industry. The superior health status of the New Zealand poultry flock is recognised all over the world. Strict hygienic and quarantine regulations are imposed on all farms.

Chicken meat is today the most-consumed meat in New Zealand. It overtook beef as the country's number one animal protein in the late 1990s and its consumption has increased dramatically since then. This change, driven by health concerns, has taken place at the expense of the red meat industries. Today, the average New Zealander eats close to 40 kg of poultry meat each year — over twice the amount consumed 20 years ago (Table 7.1).

The popularity of chicken meat is attributable to its low fat content compared with other meats, along with its versatility and convenience. It is a white meat, considered healthy, and fits into all types of cuisines. Chicken meat has no religious or cultural constraints, is easy to prepare and is less likely to be ruined by overcooking than other meats. It has become a kitchen favourite for consumers who are both pressed for time and somewhat inexpert at cooking. Price is another factor contributing to the rise of chicken meat's popularity. In constant dollars, the wholesale price of a whole chicken dropped 50 per cent from 1978 to 2000, while that of skinless, boneless breast dropped 70 per cent.

Table 7.1 **Trends in per capita consumption of beef, chicken meat and eggs**

Year	Beef (kg)	Chicken meat (kg)	Eggs (number)
1950	40	1	30
1960	46	4	60
1970	48	11	100
1980	30	17	120
1990	30	25	160
2000	30	32	200
2010	28	38	220

Collated from various sources.

Broiler farming is attractive because of its short production cycle (only five weeks) and simple management. As long as adequate nutrition is provided, the bedding in the barn is maintained dry and the shed environment is comfortable, the birds will grow fast and be efficient in converting feed into meat. In addition, the high meat yield (75 per cent of liveweight) makes broiler farming profitable.

Industry structure

The New Zealand broiler industry is characterised by a small number of large integrated companies, a position increasingly mirrored on a global scale. Four major broiler companies — Tegel Foods Ltd, Inghams Enterprises (New Zealand) Pty Ltd, P.H. van den Brink Ltd and Turks Poultry Farm Ltd — dominate the industry; between them, they supply 92 per cent of the market. Tegel is the leading chicken-meat producer, with a market share of around 50 per cent. As in most developed countries, the industry operates through vertical integration, with company ownership of all segments of the production line — feed mills, breeding farms, hatcheries, rearing farms, processing plants, marketing, veterinary services and transport. This structure enables more control over the value chain and gives companies access to greater production inputs, distribution resources, and process and retail channels.

Broilers are reared by producers under contract to one of the four broiler companies. These contract growers own the farm and provide the management, housing, equipment, labour, bedding and other inputs for the rearing of the chickens. The processing company provides the chicks, feed, medication and technical advice. The growers are paid a growing fee per bird sent for processing, with the payment also including a performance-based component. In 2015, approximately 200 contract growers produced just over 110 million meat birds (Brooks, 2016).

Bird type

The type of bird used for meat production has changed dramatically over the years. During the early days, specialised meat breeds (Cornish, Orpington) and dual-purpose breeds (Australorp, Rhode Island Red, New Hampshire, Plymouth Rock) with a blocky body conformation for optimal muscle deposition were used. The growth rate of these birds was slow; it usually took over 12 weeks to reach a market weight of 2 kg.

As the demand for chicken meat increased and the benefits of hybrid vigour (where performance of the hybrid is superior to that of either of the parents) became recognised, breeders initiated crossbreeding programmes and the use of hybrids became commonplace during the middle of the twentieth century. Since the 1970s, the need to increase meat production to meet consumer demand led to the importation of broiler strains, which have been developed by international breeding companies through selective breeding for faster growth and improved feed efficiency (Cooper-Blanks, 1999). Today, breed names have become less relevant in commercial poultry production. Ross and Cobb (the names referring to the breeding companies involved) are the two genetic strains used by the New Zealand broiler industry. These strains can reach a body weight of 2 kg in five weeks.

Housing and management

Broiler production in New Zealand comes under the provision of the Animal Welfare (Meat Chickens) Code of Welfare 2012 (NAWAC, 2012b), which sets out the minimum standards to be maintained. A typical shed on a contract farm would house 30,000–40,000 birds at a given time-point, and up to eight flocks may be grown per year. Broilers are housed in a barn with bedding on the floor (also known as a deep litter system) and are able to move freely inside the building. Stocking density is the one welfare issue of note with broiler farming; the number of birds that can be housed per unit area is dictated by the welfare code. Chickens must be managed at a stocking density that takes account of growth rate. According to the welfare code, stocking density in sheds must not exceed 38 kg of liveweight (or 20–22 birds) per square metre of floor space.

Prior to the arrival of chicks, the shed is cleaned out and disinfected, and each batch starts with new litter. Wood shavings are the most commonly used litter material. In the broiler house, proper ventilation is critical (i) for temperature control, (ii) to remove moisture and ammonia, (iii) to supply fresh air and (iv) to keep the litter dry. Maintenance of a dry and friable litter is a crucial part of broiler welfare. In addition to causing poor performance, poor air quality, diseases and carcass downgrading, a wet litter can result in hock and foot 'burns' and leg problems.

The chickens need adequate lighting to see and to locate feed and water. Lighting is controlled to provide 1–4 hours of dark period per 24 hours. The main aims of the dark period are to control over-eating by broilers, avoid sudden deaths and reduce mortality; a further aim is to prevent panic in the flock in the event of a power failure, which can cause stress and heavy mortality. The lighting provided is usually dimmer than natural light, and this tends to promote calmness in the flock. The general recommendations are 20 lux light intensity during brooding and 5–8 lux during the growing phase.

The production phase starts with the arrival of day-old chicks at the farm from the hatchery. The chicks are transported to the farms in ventilated chick boxes and in air-conditioned trucks that are specifically designed to carry chicks. It is important that the chicks receive feed, water and warmth soon after hatching, although the remains of the yolk sac can sustain them for up to 72 hours after hatching. Sometimes the sex of the chicks is determined and the males and females are reared in separate sheds, but usually they are reared together. In other countries chicks are vaccinated at the hatchery against several diseases, but in New Zealand no vaccination is carried out for broilers because of the absence of major serious diseases.

Prior to the arrival of the chicks, the shed must be heated to at least 32°C. Initially, the chicks are confined to an area between one-third and a half of the total house (the 'brooding area') and are given supplementary heating via gas heaters or heat lamps (referred to as brooders). Chicks have soft down feathers that do not protect them from low temperatures. The provision of warmth is therefore imperative until the down feathers are replaced with normal feathers; this transformation usually takes three weeks and this period is referred to as the brooding period.

The air temperature in the brooder is held at 35°C on day 1 and gradually reduced to 23°C by week 3. On most farms, regular checks are made of shed temperature, humidity and ventilation, with these factors being monitored automatically and controlled by computer. In New Zealand, broiler sheds are totally enclosed and the shed environment fully controlled. The flock is checked at least once daily to monitor bird health, to remove any dead birds and to cull any sick or injured birds. The average mortality in a well-run broiler farm over the broiler production cycle of 35 days is usually less than 4 per cent.

Feeding

Feed constitutes the major variable production cost, representing 70 per cent of the cost of producing broilers, and skilful management of feed formulation is therefore necessary to maximise profitability. Water is the most important nutrient required for animal growth, and a supply of clean water must be ensured at all times.

Because of their fast growth rate, broilers must be provided with various nutrients not only in adequate amounts but also in the correct ratios. Broilers are extremely sensitive to poor nutrition and respond immediately to any nutritional changes. In New Zealand, a feeding system involving up to five phases is typically used — pre-starter, starter, grower, finisher and withdrawal feeds. This system is referred to as phase feeding and is based on the principle that the protein and amino acid requirements of the broilers are reduced as the birds grow. The pre-starter feed is in crumble form, while the other feeds are fed as pellets. Details of the nutrient requirements of broilers during various growth phases will not be provided here. The formulations of the feeds used in the industry closely follow the breeding company guidelines (Cobb, 2014; Ross, 2014), but each New Zealand broiler company will have their own nutrient specifications based on their field experience and available research findings.

Broiler feeds are pelleted at temperatures above 85°C to ensure that the feeds are salmonella-free. Low-energy-density feeds are used because of the high cost of grains, but relatively high dietary concentrations of amino acids (the building blocks of proteins), especially lysine, are used in diet formulations. The concepts of digestible amino acids and ideal protein, with continuous refinement, form the basis of diet formulations.

Poultry diets contain a mixture of several ingredients, including cereal grains, cereal by-products, fats, plant protein sources, animal by-products, vitamin and mineral supplements, synthetic amino acids and feed additives (Table 7.2). These are invariably assembled on a least-cost basis using ingredient prices together with

Table 7.2 **An example of ingredient mixture used in broiler feed formulations**

Ingredient	Inclusion level (%)
Wheat	50–60
Barley	10
Soybean meal	20–25
Meat and bone meal	3–5
Tallow/soybean oil	2–4
Limestone	1
Dicalcium phosphate	1
Minor ingredients[1]	1–2

1. Salt, 0.2%; sodium bicarbonate, 0.2%; synthetic amino acids (DL-methionine, L-threonine and L-lysine), 0.5%; trace mineral and vitamin pre-mix, 0.2%; and various feed additives. For a list of feed additives, see Table 7.3.

nutrient contents. The major feed ingredients are wheat, soybean meal, and meat and bone meal (MBM). In recent years, increasing amounts of barley has been used in broiler diets. Meat and bone meal, with its relatively high levels of protein, calcium and available phosphorus, is a valuable local protein source.

However, it is a highly variable product in terms of the content and digestibility of amino acids. Feed manufacturers nowadays also have to cope with increasing safety concerns, epitomised by the crisis arising from bovine spongiform encephalopathy (BSE; commonly known as mad cow disease) associated with the feeding of contaminated MBM. Although BSE did not occur in New Zealand, the use of MBM in animal feed manufacture is now banned in some parts of the world and the long-term future of this raw material in New Zealand may thus be uncertain.

Current domestic wheat production does not always meet the requirements of the feed industry. To meet this shortfall, sorghum is imported from time to time from Australia. Maize production is limited and often erratic; in consequence, only limited amounts of maize are available for feed manufacture. The formulations usually contain 2–4 per cent added fats, usually as a mixture of tallow and vegetable oil. Currently there are no locally produced plant protein sources, and all soybean meal demand is met by imports. Soybean meal is an important feed ingredient as a primary source of protein and a secondary source of energy. The greatest single challenge faced by the New Zealand broiler industry relates to the availability and cost of feed ingredients. Of major concern are also the wide variability in the quality of locally produced ingredients (wheat and MBM), and the

Table 7.3 **Feed additives commonly used in poultry feed formulations**

Additive	Examples	Reasons for use
Enzymes	Xylanases, β-glucanases, phytase	To overcome the anti-nutritional effects of arabinoxylans (in wheat and triticale), β-glucans (in barley) or phytate (in all plant feedstuffs) and to improve the overall nutrient availability and feed value
Antibiotics[1]	Avilamycin, zinc bacitracin, flavomycin	To control harmful bacterial species in the gut and to improve production efficiency; as a prophylactic measure against necrotic enteritis
Coccidiostats	Monensin, salinomycin, narasin	To prevent and control the clinical symptoms of coccidiosis
Pigmentants	Xanthophyll (natural and synthetic)	To improve the skin colour and appearance of carcasses
Antioxidants	Butylated hydroxytoluene, butylated hydroxyanisole, ethoxyquin	To prevent auto-oxidation of fats and oils in the diets (this makes the feed rancid)
Antifungals		To control mould growth in feed; to bind and mitigate the negative effects of fungal toxins
Antibiotic replacers[2]		
Direct-fed microbials	Probiotics	Provision of beneficial species such as lactobacilli and streptococci
Prebiotics	Fructo-oligosaccharides, mannan-oligosaccharides	Binding of harmful bacteria
Organic acids	Propionic acid, diformate	Lowering of gut pH and prevention of the growth of harmful bacteria
Botanicals	Herbs, spices, plant extracts, essential oils	Prevention of the growth of harmful bacteria

1. The use of other in-feed antibiotics has been voluntarily withdrawn by the New Zealand poultry industry since 2001.
2. In anticipation of a total ban on in-feed antibiotic use, a multitude of these compounds (individually and in combination) are currently being tested. Source: Ravindran (2013).

non-availability of locally grown plant protein feeds that can replace imported soybean meal.

Poultry formulations also contain an array of substances known as feed additives. These are non-nutritive substances usually added in amounts of less than 0.05 per cent to maintain health status, uniformity and production efficiency in intensive production systems. A list of commonly used feed additives is presented in Table 7.3. Since wheat is the major grain used in poultry feeds (Table 7.2), xylanases to aid digestion are routinely included in all diets. Despite the high dietary levels of available phosphorus (from the use of 4–8 per cent MBM in formulations), phytase enzyme is also used. Coccidiostat, a medication, is added to broiler feeds to prevent the intestinal disease coccidiosis (caused by various species of protozoa). These additives have now become vital components in poultry diets.

Production systems

Commercially, broilers are housed on a litter floor in sheds (or barns), and not in cages as some consumers believe. Free-range broiler farming has received some attention in recent years due to consumer demand. In this system, birds are housed in a barn but can roam freely outdoors for at least part of the day. The birds have access to natural conditions and can exhibit their normal behaviour. Currently, 97 per cent of meat chickens grown in New Zealand are produced in barns and the balance in a free-range system.

Table 7.4 **Trends in the production parameters of male broilers in New Zealand, 1965–2015**

Year	Days taken to reach 2 kg body weight	Feed conversion ratio (unit feed/unit body weight gain) to 2 kg
1965	84	3.50
1975	63	2.70
1985	45	2.30
1995	40	2.00
2005	35	1.60
2015	30	1.45

Productivity levels

Over time, bird growth has dramatically improved while the amount of feed required per unit growth has steadily declined. The time required for male broilers to reach a body weight of 2 kg has declined from 40 days in 1995 to 30 days in 2015. During the same period, the feed conversion ratio (unit of feed consumed per unit of body weight gain) at the farm level has decreased from 2.00 to around 1.45 (Table 7.4). More efficient farms are known to realise conversions of less than 1.40:1. Under optimum nutritional and husbandry conditions at Massey University's Poultry Research Unit, conversions of 1.22:1 have been achieved. These performance improvements are due largely to genetic gains achieved over this period through conventional selective breeding. Vast advances in nutritional knowledge are the second most important contributing factor. Other factors, including husbandry techniques and health management, have also contributed.

The superior feed conversion ratios achieved in New Zealand compared with other countries (Table 7.5) can be attributed largely to a minimal immune challenge resulting from the better health status, better hygiene conditions and lack of any vaccination of broiler birds. The industry continues to work towards a steady reduction in feed conversion ratio by identifying factors that contribute to performance losses. Strategies being addressed include, but are not limited to, nutrition of the chick during the first 10 days (nutrient-dense, highly digestible feeds, supplementation with specific amino acids, specific feed additives), coccidiosis control, water quality, shed environment (temperature and air quality), refinement of ideal amino acid

Table 7.5 **Comparison of production parameters of male broilers in New Zealand versus other countries**

Country	Days taken to reach 2 kg body weight	Feed conversion ratio (unit feed/unit body weight gain) to 2 kg
United States	35–40	1.80
United Kingdom	38–40	1.80
China	40–45	2.00
Australia	33	1.60
New Zealand	30	1.45

ratios (relative to lysine) during different growth phases, and expansion of databases on digestible amino acids.

Marketing

Chicken meat in New Zealand is sold under the brand name of the company, and market competition is very strong. The time of marketing is determined by liveweight and the requirements of retailers. Birds may be harvested (also known as partial depopulation or thinning out) at up to four different times. Fast-food outlets, such as Kentucky Fried Chicken, require a smaller bird weighing 1.8 kg liveweight and giving a dressed weight of around 1.25 kg. The typical supermarket broiler, harvested at a liveweight of 2 kg, has a dressed weight of 1.55 kg. When special carcass cuts, especially breast meat, are of interest, the birds may be grown to up to 4 kg liveweight. Harvesting allows more space for the remaining birds and reduces the ambient temperature in the shed. The first harvest takes place at around 28–33 days of age and the last at 55–60 days.

The type of meat marketed in New Zealand has changed dramatically over the past 50 years to meet the demands of consumers. In the 1960s, chicken meat was available only as frozen whole birds. Today, only 30 per cent is sold as frozen products and another 30 per cent as fresh; about one-third of the meat is sold as whole birds and the balance as speciality cuts, fillets and value-added products (Table 7.6). When whole birds are sold, the carcasses are given a size number corresponding to the weight of dressed bird. For example, whole chickens with dressed weights between 1.25 and 1.35 kg are sold in New Zealand supermarkets as size 13, and those between 1.55 and 1.65 kg as size 16.

During harvesting, birds are removed from the farm at night when they are quieter; this also ensures that they arrive at the processing plant early in the morning with less delay before slaughter. Any delay will result in stress and weight loss. Birds are placed into crates designed for good ventilation and safety from bruising during transport. The transport and slaughter of broilers is regulated by the Animal Welfare (Transport within New Zealand) Code of Welfare 2011 and the Animal Welfare (Commercial Slaughter) Code of Welfare 2010 (NAWAC, 2010; 2011).

Table 7.6 **Trends in the marketing of chicken meat products, 1960–2000**

Decade	Products
1960s	Frozen whole birds
1970s	Fresh whole birds
1980s	Portions: fresh and frozen Further processed: cooked and uncooked
1990s	Food service: pre-prepared restaurant items, further processed items available
2000s	Convenience foods: pre-cooked, fast-cook, easy-cook products Speciality products such as free-range and organic birds

Source: Brooks (n.d.).

The carcasses are graded during the processing sequence to remove poor-quality meat. This meat is used for cut-up (further processing) purposes or, if badly affected, might be used for pet food, or condemned and cooked to be made into poultry by-product meal for use in pig feeds. Further processing includes cutting up the carcasses into portions, deboning carcasses and preparing special ready-to-cook products. Almost all chicken meat is sold chilled. The shelf-life of chicken meat is usually 8–12 days, depending on the processing, handling and storage conditions.

Broiler chickens outdoors.

Once all of the birds have been harvested, the shed is cleaned and prepared for the next batch of chicks, which generally arrives 5–10 days after the harvest. The time between batches allows for the cleaning and disinfection of the shed, and reduces the risk of pathogenic organisms being transferred between batches.

Consumer issues

Welfare, food safety and environmental concerns dominate a large part of discussions associated with poultry production. However, compared with the layer industry the welfare issues in the broiler industry are minor; the main concern being stocking density.

The possible link between food-borne diseases and chicken meat is an important consumer issue that has been the subject of considerable negative media reporting over the past decade. The industry has always recognised that food and feed safety are areas that need continuing vigilance and improvement, and has been proactive in investing heavily in quality assurance and biosecurity programmes.

Two food-borne diseases, namely salmonella and campylobacteriosis, are particular concerns in the poultry industry. In chicken meat the salmonella issue has been brought under control by regular testing and stringent management. A survey carried out in 2010 to determine the prevalence of salmonella on retail chicken meat products found that the incidence was less than 1 per cent (Brooks, 2016). In contrast, a high and increasing burden of campylobacteriosis in the general population was reported in 2006 (Baker et al., 2006b). The rate of notified pathogenic campylobacter cases (per 100,000 population) in New Zealand was high (396) compared with

A typical chicken shed.

those reported for other developed countries (Australia, 116; United Kingdom, 86; United States, 13; Canada, 40).

While no major cause for this trend has been conclusively identified, fingers are often pointed at the overall increase in poultry consumption through the years (Baker et al., 2006a; Duncan, 2014). Today, however, due to a range of food safety and industry interventions implemented since 2006, a 74 per cent decline in campylobacteriosis originating from poultry sources has been reported (Sears et al., 2011).

Due to consumer concerns, in 2001 the poultry industry voluntarily withdrew the use of most in-feed antibiotics in broiler diets. Currently, the use of only three in-feed antibiotics (avilamycin, zinc bacitracin and flavomycin) is allowed, mainly as a prophylactic measure against *Clostridium perfringens* (which causes a fatal disease known as necrotic enteritis). In anticipation of a future ban on the use of in-feed medications, the industry has invested in research into the study of gut health and the evaluation of a wide range of alternatives currently available on the market. Strategies that involve the use of organic acids, probiotics, prebiotics, feed enzymes and plant extracts are being evaluated.

The focus of these alternative strategies has been to prevent the proliferation of pathogenic bacteria, thus maintaining health, immune status and performance, and optimising digestion in poultry production. It is being increasingly recognised that stable gut microflora, predominantly composed of lactic-acid-producing bacteria and protecting the host from pathogenic invasion, is a prerequisite for gut health and good growth performance.

Day-old broiler chicks.

Chicken meat — urban myths

The reputation of chicken meat suffers from several misconceptions. One myth circulating is that growth hormones are used in the production of broiler chickens, and therefore the chicken meat contains hormones. There are two reasons for this myth. First, although in some parts of the world oestrogens were indeed administered to male layer-type chicks to increase growth until the 1960s, this was never practised in New Zealand. No hormones are fed or otherwise administered to poultry in this country.

Second, the increasing size of chickens and breast meat cuts over the decades has, naturally, raised concerns about the possible use of hormones. As noted above, however, the increased size is largely due to genetic improvements in broilers and not to the feeding of hormones. Contemporary strains of meat chicken have been highly selected, with their genetic potential altered to grow twice as fast and twice as large as their counterparts a few decades ago.

Related fallacies are that eating too much chicken meat can cause larger breasts in humans and that pregnant women should not eat chicken meat. This is again based on the hormone myth and, as noted above, that has no foundation. In fact, chicken meat is the only animal protein in New Zealand that is routinely tested for the presence of growth hormones and none has ever been found.

Another misconception is that chicken meat contains antibiotic residues, and that this can cause adverse reactions in consumers by building resistance against antibiotics used in humans. It is true that certain types of antibiotics used in human medicine have also been used in broiler feed in the past. However, the antibiotics currently used in broiler diets are not used in human medicine and are also not absorbed by the gut so do not leave any residues in the meat. Additionally, antibiotics are withdrawn from feed 7–10 days before slaughter, allowing time for any residues to be removed naturally from tissues. There should therefore be no concern about the presence of antibiotic residues in chicken meat available in New Zealand.

Turkey and duck meat production

A brief mention must be made of the other types of commercial poultry — turkeys and ducks. Meat production from these two species represents only a very small portion of that of commercial chickens. Exact statistics on the numbers of turkeys and ducks produced in New Zealand are not available.

Turkey operations are located only in the South Island. One major company (Tegel) and several independents are involved. Three batches of turkeys are produced per year, and the total annual output is thought to be around 500,000 birds. Over 90 per cent of the output is sold frozen during the Christmas period. Unlike the heavy turkeys grown to a body weight of 16 kg in the United States, the birds grown in New Zealand are broiler turkeys. Broiler turkeys reach a body weight of 5–7 kg in 12–14 weeks, with a dressed weight of 3.5–5 kg.

Duck meat operations are spread throughout the country, around major cities where the meat caters to the demands of ethnic communities and speciality restaurants. Duck meat production is a rapidly growing area. The farms are independently operated and of small scale. The annual output is thought to be around 200,000 birds. Ducks are generally grown for 6–8 weeks, reaching a market weight of 3.5 kg.

Table egg production

The value of eggs as a food for humans has been known from prehistoric times. Humans' first experience with eggs was probably through gathering them from the nests of wild birds and they would have been considered a delicacy. Following the domestication of chickens 10,000 years ago, the supply of eggs became regular. Today, with improvements in breeding and husbandry, a hen can produce 300–320 eggs per year and eggs have become a year-round staple.

Nutritional value of eggs

The egg is an inexpensive and highly digestible source of protein. The amino acid content of an egg is so well balanced that the egg is considered to be the 'gold standard' against which other proteins are compared. Eggs contain every vitamin except vitamin C. They are particularly high in vitamins A, D and B12, and also contain B1 and riboflavin. Eggs are second only to fish liver oil as a natural source of vitamin D. They are a good source of iron and phosphorus and also supply calcium, copper, iodine, magnesium, manganese, potassium, sodium, zinc, chloride and sulphur.

All of these minerals are present as organic chelates and are therefore highly bioavailable. The yellow pigments lutein and zeaxanthin, which are responsible for the colour of the yolk, are antioxidants. These are very efficiently absorbed by humans and highly available for the maintenance of retinal health. All of these features suggest that egg is the most complete whole food in terms of nutrition.

Compared with other animal products, the

fat and saturated fatty acid contents in eggs are not high and the egg fat is predominantly unsaturated. For years, the image of the egg has been tarnished as being unhealthy due to its cholesterol content and it has often been blamed for an increased risk of heart disease. As a result, despite their taste, value, convenience and nutrition, consumers have avoided eating more than a few eggs for fear of health issues. According to current nutritional data, however, eggs are lower in cholesterol than previously recorded; results show that the cholesterol content in one medium-size egg is 180 mg — as opposed to the 227 mg reported in the 1960s.

The egg is a good source of healthy fat, and research has documented that healthy adults can enjoy an egg a day without significantly affecting their risk of heart disease. Dietary guidelines now recommend that individuals consume, on average, less than 300 mg of cholesterol per day and, as noted above, an egg contains only 180 mg cholesterol. Today the important nutritional contribution of the egg is widely acknowledged and it is being promoted as part of a heart-healthy diet.

Salmonella is a food-borne disease commonly associated with egg consumption in most countries. In New Zealand, concerted efforts by the industry have resulted in salmonella contamination being almost non-existent in eggs. This bacterium has never been detected inside an egg in New Zealand (Brooks, 2016).

Currently the egg is also being viewed with interest, beyond its traditional food value, for its biologically active components that are naturally present for the defence of the developing embryo. These components are unique and have an array of novel food and non-food applications (Ravindran, 2012).

Industry structure

The table egg industry operates both through horizontal and vertical integration. Two-thirds of the industry is dominated by independent producers who operate via horizontal integration. These individual farms have no ownership over various components of the production process, all buying day-old chicks from commercial hatcheries and most buying feed from commercial feed compounders. Many producers sell direct to the wholesale or retail trade, while some producers market through a co-operative (Independent Egg Producers Co-operative Ltd) under a common brand, Morning Fresh.

The weakness of this horizontal integration is that producers have little control over the market, which often results in fluctuating supply and price of eggs. The size of egg farms has increased and farm numbers have declined over the years; currently, only 125 farms are involved in the production of table eggs as compared with 633 in 1980. Since 1997, a vertically integrated company (Mainland Poultry Ltd), based in Dunedin, has been involved in table egg production and maintains around 1 million layers. The eggs are marketed under the brand name Zeagold. The egg market is currently dominated by Mainland Poultry Ltd and Independent Egg Producers Co-operative Ltd, which together account for two-thirds of total egg production.

Domestic egg production grew by 5.8 per cent between 2014 and 2015 and is about 88 million dozens per annum, which comes from a national layer flock of 3 million birds. An average New Zealander consumes about 225 eggs per annum and about 20 per cent of egg production goes into the manufacture of liquid egg products (frozen or pasteurised) and dried egg white. Egg products are used extensively in the food manufacturing industry owing to their useful functional properties such as foaming ability, binding ability, emulsification, colour and aroma.

Bird type

Similar to the progress made in broiler genetics, the bird type used for table egg production has changed over the years. The White Leghorn (WL), with an average annual egg production of 220 eggs, was the breed of choice during the early days. Dual-purpose breeds (Australorp [AO], Rhode Island Red [RIR] and Plymouth Rock [PR]) were also used; these birds were reasonable layers (180 eggs per annum), with the added advantage of meat supply at the end of the laying period.

By the 1960s, producers started using crossbred birds, predominantly WL x AO crosses but also WL x RIR and WL x PR; these crossbreds being multiplied throughout the country by a network of small breeders. During the early 1970s, import restrictions on stock from the United Kingdom and Canada were lifted as a result of improvements in disease diagnosis and control. The imported egg strains outperformed the local layer breeds and became a permanent feature by the mid-1970s.

The layer strains currently used by the industry are the Hyline (both brown and white eggs) and the Shaver (brown eggs only). These strains can produce over 300 eggs per hen in 52 weeks and some hens are known to lay over 340 eggs — almost an egg every day. The production improvements, summarised in Table 7.7, reflect both the genetic changes over the years and advances in scientific knowledge relating to nutritional requirements and bird management.

Table 7.7 **Trends in production parameters of laying hens, 1960–2015**

Year	Egg number per 52 weeks	Feed conversion ratio (unit feed/unit egg weight gain)
1960	190	4.00
1970	240	3.50
1980	270	2.80
1990	290	2.50
2000	300	2.30
2015	> 300	2.20

In the 1920s, poultry farmers were content with an egg output of below 100 per hen (Ambler, 1923).

Management of layers

Chick to pullet stage

The period from one day old to 17 weeks post-hatching is the most critical time of a laying bird's life. Mistakes made during this time are very difficult to overcome. The major goal for the pullet-growing programme is to produce high-quality birds that will achieve the optimum possible laying performance. Achieving the target weight for age and good uniformity are means to achieving that goal. The same general principles of husbandry for broiler chickens apply to layer chickens, but there are major differences in the feeding, lighting and vaccination programmes.

Lighting programme

Lighting plays an important role in sexual maturity and egg-laying in poultry. Day-old chicks should start with at least two days of continuous light at an intensity of 10 lux, after which the light should be reduced to 15 hours per day at 5–10 lux intensity. From three to 18 weeks of age, a constant day length of 10–12 hours must be maintained to avoid any delay in maturity. The golden rule is never to increase day length for growing pullets prior to week 18 of life.

Sexual maturity is advanced by increasing the photoperiod (period of increased light), which stimulates hormonal production and the development of the reproductive tract. When the target body weight is reached, the light stimulation is increased by 15–30 minutes per week until peak production at 30–32 weeks. Light intensity should also be gradually increased from 10–20 lux.

Vaccination programme

Vaccination is an important way of preventing diseases. Attenuated vaccines are administered to birds by different routes: drinking water, eye drops, spray, subcutaneous injection, intramuscular injection or wing stab. Table 7.8 gives a suggested programme to be used in New Zealand.

Starter phase (one day old to six weeks of age)

At the hatchery, newly hatched chicks are sexed; the females are sold as future layers while the males are humanely euthanased. The gender of a day-old chick is determined by cloacal examination, feather colour or feather length. When the female chicks arrive at the farm, the rearing shed must have been cleaned, disinfected and pre-heated. Chicks may be reared in the layer farm itself or in a replacement pullet farm, which grows and supplies 16-week-old point-of-lay pullets.

For the first 3–6 weeks, the chicks are brooded under heat — similar to the brooding management of broilers. The brooding period may last up to six weeks, depending on the temperature of the environment, until the chicks develop normal feathers to control their body temperature themselves. The brooder shed is usually located some distance away from older flocks for reasons of hygiene and to prevent the transmission of disease.

Table 7.8 **Example of a vaccination programme for layer-type birds, from hatching to 15 weeks of age**

Time	Route	Vaccination
Hatching	Spray	Marek's and salmonella
2 weeks	Spray	Salmonella
5 weeks	Spray/drinking water	Infectious bronchitis
10 weeks	Wing stab	Avian encephalomyelitis
15 weeks	Spray/drinking water Subcutaneous injection Intramuscular injection	Infectious bronchitis Egg drop syndrome (EDS 76), *Mycoplasma galliseptum*
15 weeks (in Auckland only)	Eye drop Wing stab	Infectious laryngotracheitis (ILT) Fowl pox

There are two systems for brooding egg-type chicks:

- **Floor brooding:** the chicks either brood for up to six weeks on the litter and are then moved to cages, or stay on the litter until transferred to layer cages at 16 weeks of age.
- **Cage brooding:** each brooder cage can hold up to 35 chicks up to four weeks of age, or up to 8–18 pullets up to 16 weeks when they are transferred to layer cages.

At 5–6 days of age, the beaks of the chicks are trimmed — an important management practice to reduce the risk of damage from feather pecking or cannibalism, reduce feed wastage, and for general calming effects in the flock. At this age, debeaking is bloodless and easily managed. It is carried out using electric debeakers and a cherry-red blade at between 595°C and 700°C has been recommended for proper cautery. According to the animal welfare code, the operator should not remove more than one-third of the upper or lower beaks.

The starter period is the most efficient period of growth. The organs, immune system and skeleton are developed by six weeks of age, and with a good foundation the pullet can grow out to a highly productive bird. A chick starter feed (18 per cent protein), which is well balanced in all of the essential amino acids and energy, is fed. The aim of feeding and management is to ensure that the target weight recommended by the breeding company is being achieved.

Body weights should be monitored periodically to determine flock average weight and uniformity. At least 100 birds should be weighed and this sample must be representative of the flock. Birds should be individually weighed each week starting at three weeks of age. Factors that can affect body weight and uniformity are the weight of day-old chicks, housing conditions, stocking density, disease, the quality of beak trimming, and nutrient intake.

Growing phase (seven to 15 weeks of age)

The mature size of the bird is determined during this period. The pullet normally achieves its optimum frame size by 12–14 weeks of age. It is known that 90 per cent of the frame size is developed by this stage and that the 'size' of the pullet is then fixed. Another important aim is to delay sexual maturity until 18–22 weeks of age to avoid an early start to egg production. Contemporary layer strains are precocious and are capable of starting to lay early unless their growth is kept under control. If a bird starts producing eggs early, then bone calcium reserves will be low and the bird will be susceptible to bone breakage and diseases. The egg size will be smaller, the laying period will be shorter and the overall productivity will be affected.

Early lay is prevented by feeding a pullet grower diet, which contains less protein (15–16 per cent) and energy than the chick starter diet. Energy intake is the major nutrient controlling body weight during this period. To control body fat deposition, energy content must be restricted after 14 weeks of age.

According to the welfare code for layers, the amount of space for pullets (birds of 7–18 weeks of age) must:

- be a minimum of 370 cm^2 per pullet for those reared in cages or colony cages, and
- not exceed 14 pullets per m^2 for those reared in barns.

Pre-lay phase (16 weeks of age to first egg)

Pullets are usually moved into their laying

quarters at 16–18 weeks of age, before they start laying. Before the transfer, the cages and house should be cleaned and sanitised. Transfer to the laying quarters is a major source of stress. It involves changes both in environment and equipment and in development of the reproductive organs (ovary and oviduct), which occurs during the 10 days prior to the first egg being laid. Handling during the move must be done with care to avoid injury. A late transfer often leads to a delayed start of lay and higher mortality during the production period.

A special pre-lay diet, with high calcium content (2 per cent), is fed during this period to build up bone calcium reserves. During egg production, calcium used for the formation of egg shell comes from both the diet and the bones, and sufficient bone calcium reserves are essential for good shell quality.

Laying period (20–78 weeks)

Housing systems for layers

Three systems of housing, namely cages (conventional and colony), barn and free-range systems, are used in layer farming. It is estimated that 81 per cent of the hens in New Zealand are housed in cages, 14 per cent in free-range systems and 5 per cent in barns (Brooks, 2016).

Cages

(i) Conventional cages

Cage housing of laying hens is probably the most criticised animal agriculture practice. Conventional cages do not provide hens with sufficient space, nest boxes or perches. Lack of space means that the hens are not able to walk, flap their wings, perch or make a nest and it is widely considered that hens suffer from boredom and frustration through being unable to perform these behaviours.

In this system, birds are confined to small cages, each housing 4–8 hens. The floor is sloped wire mesh that allows the excreta to drop through and the eggs to roll out. Water is usually provided in overhead nipple drinkers, and feed troughs along the front of the cage are replenished at regular intervals by a mechanical chain. The cages are arranged in long rows as multiple tiers, often with cages back-to-back. Within a single shed there may be several tiers of cages, and a single shed may contain several tens of thousands of hens. Light intensity is often kept low (e.g. 10 lux) to reduce feather pecking and vent pecking.

The industry moved to indoor cage housing in the 1960s to meet the increasing demand for eggs. Cages allowed more birds to be housed per unit area, enabled greater productivity and lowered the prices. Other benefits included more control over management, easier care of the birds, lower mortality and reduced labour requirements.

The European Union banned the use of conventional battery cages from January 2012. Around 80 per cent of New Zealand's 3 million laying hens currently live in battery cages. The current welfare code for layers (NAWAC, 2012a) recommends that, as of 1 January 2014, such cages provide a minimum of 500 cm^2 of unrestricted floor space per hen. Conventional cages will be phased out in New Zealand by 2022 and the establishment of new conventional cage farms is currently not permitted.

(ii) Colony cages

Colony cages (also known as enriched or furnished cages) are designed to overcome the major welfare concerns regarding battery cages while retaining their economic and husbandry advantages, and also provide some of the welfare advantages of non-cage systems. These cages provide perches, a nest, a litter area and more space per bird than conventional cages. The welfare code for layers recommends a minimum stocking density of 750 cm^2 per hen (or 13 hens/m^2) in colony cages. The egg production and mortality of hens in colony cages are reported to be similar to those in conventional cages.

Deep litter (or barn) system

The deep litter house or barn is exactly the same as the broiler house. In this system, wood shavings are placed on a concrete base. In the barns, hens are kept at a single level on litter. A typical design would have nest boxes arranged on slats in the centre of the house. The welfare code for layers recommends that stocking density should not exceed seven hens/m^2 in barns. Litter management is critical and should be checked frequently for dryness and friability. If litter becomes wet or caked, the problem should be immediately rectified.

Aviaries are essentially a barn system, but with multi-tiers of platforms at different levels which enables a high stocking density to be achieved. Aviaries may contain extensive perches, thus making use of the height of the house. These systems are known as percheries; they usually provide less floor space per bird but compensate for this by providing perches arranged on frames. This feature enables the movement of hens up and down between perch levels. Feeders and drinkers are usually located only on the bottom and middle levels. A stocking density of up to 14 birds/m^2 of floor area should not be exceeded in percheries. A wide range of perchery options are now commercially available.

Free-range system

There is no agreement on what exactly constitutes free-range poultry, but it may be defined as a system that provides the birds with access to an extensive outdoor area and which typically includes an indoor shelter (fixed or portable) for the night or poor weather. Stocking density in the outdoor ranging area, as specified in the code of welfare for layers, must not exceed 2500 hens per hectare.

The benefits of free-range farming include opportunities for hens to exhibit natural behaviours such as pecking, scratching and foraging, and exercise in the outdoors. The disadvantages include land requirement, loss of control over management and, potentially, high mortality from parasites, diseases, inclement weather and predators.

Organic egg production

This describes a production philosophy rather than a housing system. Organic farming is defined as farming without the addition of artificial chemicals. In organic egg production (or broiler) systems, birds may be kept in any housing type but feeding is based on restrictions on the use of synthetic feed additives such as vitamins, mineral sources, medications and amino acids. Ingredients used in feed (e.g. grains) should come from farms where no chemicals (pesticides, fungicides, weedicides or inorganic fertilisers) have been applied.

Under the regulations, no antibiotic medication is to be used for the treatment of organic poultry and the birds cannot be routinely vaccinated. There are exceptions, however, such as where treatment is required by law or disease cannot be controlled with organic management practices.

Lighting programme

Three lighting programmes, namely constant, step-up or step-down, are used in layer houses, depending on the stage of production. Laying flocks are usually given at least 13 hours of light at an intensity of 10 lux when egg production is desired, with additional weekly increases of 15–30 minutes until a maximum of about 16 hours of total day length is reached (Table 7.9). It is important to first let pullets reach the target body weight before increasing the light.

Table 7.9 **An example of a typical lighting programme for layers, 16–32 weeks of age**

Age (weeks)	Details	Photoperiod (hours)
16	Start light stimulation	10
17	Introduce pre-layer diet	11
18	First egg	12
19	15% production	13
20	Introduce layer feed	14
21	50% production	14–16
26–32	Peak production (over 90%)	14–16

Layer management

The egg is formed in the reproductive tract of the hen. Egg formation is a natural process and it is not necessary for a hen to mate with a rooster to produce an egg. Roosters are not housed with hens in any of the commercial layer production systems in New Zealand. Hence, the table eggs marketed locally are unfertilised and some consider these to be vegetarian. Egg formation begins with the release of an ovum (yolk) from the ovary. As the ovum travels down the reproductive tract, albumen, shell membranes and shell are sequentially added to form the egg. The minimum time needed for the full process (from ovum release to egg lay) is 24 hours and the next ovum is only released after the egg is laid; hence, a hen can produce only one egg per day.

Eggs laid on successive days are known as a sequence. The first egg is laid at first light, with subsequent eggs being laid later in the day until the rhythm is broken, usually around mid-afternoon. There will then be a rest period (one or more days) between sequences. The total number of eggs laid during a sequence and the length of the rest period determines the egg production profile of the flock. The current strains lay more eggs per sequence and have shorter rest periods, resulting in high productivity.

The first egg is laid at around 18 weeks of age. Egg production increases sharply during the next 12 weeks, reaching peak lay of 90–95 per cent (90–95 eggs/day/100 hens) at 30–32 weeks of age. In contemporary strains, a production rate of over 95 per cent is sustained for six months or more. Current commercial strains may lay 300-plus eggs over 52 weeks.

Egg size also increases with advancing age of pullets. The first eggs (known as pullet eggs) are small, weighing around 35 g. A large proportion of the eggs reach a weight of around 55 g by the time of peak production. Shell quality, however, declines towards the end of the laying cycle. At 78 weeks of age, egg production falls below 60 per cent with a corresponding decline in shell quality. There is a greater incidence of thin-shelled eggs and egg breakages; these eggs are not saleable and production becomes uneconomic.

This marks the end of the first production cycle; the common practice is to replace hens with new stock. Culled birds can be processed and sold as spent hens or roasters, or used in pet foods. If

Five-week-old broiler chicks.

there is no ready market, the birds are humanely euthanased and composted for use as fertiliser.

In certain situations, such as during a disease outbreak or the non-availability of replacement stock, artificially induced moulting (also known as forced moulting) may be used to extend the productive life of the flock. The aims are to get birds to stop laying quickly, reduce body weight, give a resting period of 4–6 weeks to the reproductive organs and build up bone calcium reserves. Moulting systems, if used in New Zealand, are based on light control together with withdrawal or restriction of the amount of feed and water that is given to the birds. After moulting and upon full feeding, hens return to a new peak lay in 4–6 weeks, with larger eggs and a good shell quality, and can be productive for another 40 weeks. However, induced moulting is a welfare issue and is only practised when dictated by market conditions.

For best performance, the temperature in layer houses must be maintained at around 21°C. Egg production decreases at average house temperatures of 27°C or above and egg size decreases at temperatures above 24°C.

Good record-keeping and routine culling of under-performing birds are essential management tools throughout the life cycle. During the laying period, culling is required to identify and remove non-laying or low-producing hens from a laying flock. Pullets that have not started to lay by 30 weeks of age should be removed. Culling will make more feed and space available for the productive birds. Culled birds should be euthanased humanely.

Feeding

The specific nutrient requirements of laying birds from hatching to the end of laying will not be discussed here; readers are referred to company guidelines (Hyline, 2016; Shaver, 2015) for details. Each layer farm will follow its own nutrient specifications in feed formulations, based on the guidelines.

In general, the nutrient requirements of layers depend on the nutritional needs of the birds and on feed intake. Feeding programmes assume that the amount of nutrients needed each day remains the same but the feed intake increases throughout a hen's time of production. Feed intake is controlled in birds by the energy level of the diet; as the energy level increases, the birds eat less (and vice versa). Bird type is another important factor in determining feed intake, with white egg strains consuming less feed than brown egg strains (105 vs 120 g/hen/day). When feed intake changes, nutrient specifications must be modified to maintain a certain level of nutrient intake. Other factors influencing feed intake are house temperature, rate of production, egg size and body weight. Thus, an estimation of feed intake is the most important factor in layer feeding programmes. Without accurate feed intake data, the formulation of layer feeds for proper nutrient intake is not possible.

The cycle of egg production can be divided into the following four phases: pre-lay, pre-peak, peak and post-peak. The nutritional quality of the feed provided to hens must be varied according to the level of production, and this must be cost-effective. Hens need more nutrients just before and during their peak production than at other times. In a phase feeding programme, the protein and amino acid contents of the diet are reduced as the hen ages or when egg production in the flock declines.

Table 7.10 **An example of ingredients used in the formulation of laying hen diets**

Ingredient	Inclusion level (%)
Wheat	25–30
Barley	25–30
Broll[1]	15–20
Soybean meal	10–12
Palm kernel meal/canola meal/lupins	5–8
Meat and bone meal	4–6
Tallow/soybean oil	1–3
Limestone	8–10
Dicalcium phosphate	1.5
Minor ingredients[2]	0.6

1. Wheat milling by-products are collectively known as broll in New Zealand.
2. Salt, 0.20%; synthetic amino acids (DL-methionine, L-threonine and L-lysine), 0.25%; trace mineral and vitamin pre-mix, 0.20%; and various feed additives. For a list of feed additives, see Table 7.3.

In contrast to broilers, laying hens do not have high nutritional requirements and are tolerant to wide-ranging nutritional conditions. Their diets are usually formulated with an array of cheaper, less digestible ingredients (Table 7.10) and offered in mash form. A unique feature of layer diets is their high calcium content (4 per cent, compared with around 1 per cent for other chicken diets). Modern layer strains lay an egg almost daily and require an intake of 4–5 g of calcium per day for good shell quality and to replenish bone calcium reserves. Egg shell is made of calcium carbonate, the calcium coming from both the diet and bone reserves. The rate at which calcium is removed from the blood during shell formation is greater than the rate of absorption of calcium from the diet; the balance is supplied by mobilisation of the bone reserves.

Egg collection, quality and storage

Eggs are collected as soon as possible after being laid and are held in cool storage to protect internal quality. The mechanical collection of eggs is common in larger layer farms. Most eggs are laid in the morning and should be collected regularly. In the egg room, eggs are candled (where a light source is shone through the egg) to remove those that are blemished, cracked, pimpled or thin-shelled; on average, about 5 per cent of eggs are unmarketable due to shell quality problems. A sample of eggs is also broken open to check the internal quality. The eggs are then graded mechanically using an egg grader and packed into cartons of six or 12 eggs or trays of 30 eggs for sale. Price varies with egg size; grading is done by egg weight, based on a minimum size for each grade (Table 7.11).

Table 7.11 **Egg sizes (grades) in New Zealand**

Weight classification	Egg size (grade)	Minimum egg weight (g)
Jumbo	8	68
Large	7	62
Standard	6	53
Medium[1]	5	44
Pullet[1]	4	35

1. The market for eggs smaller than size 6 is weak, and these are normally used for pulping and liquid egg products.

Consumer judgement of egg quality is based primarily on the external appearance of the egg. The consumer wants a clean egg of a certain shape with a strong shell that may vary in shade between white and brown. The colour of the shell is genetically controlled and shell colour has no influence on the nutrient value of the egg. The preference of local consumers is for brown eggs, which may be partly because of the belief that brown eggs are more nutritious and healthier. Controlled comparisons, however, have shown that there are no differences in the taste, quality or nutritional profile between these two egg types.

When broken, the egg should give an overall attractive appearance with respect to yolk colour, albumen condition, the way the egg sits up, and freedom from abnormal spots and odour. The quality of the albumen is a good indicator of the freshness of the egg. The colour of the yolk depends on what the hen eats. Free-range eggs have a darker yellow-orange yolk because of the hens' access to grass. In cage and barn systems, the colour is maintained by the inclusion of natural or artificial pigments in hen diets.

In the farm and retail outlets, eggs are stored in a cool room at about 13°C. Eggs purchased by consumers are best stored at normal refrigerator temperature (4°C). Eggs refrigerated after purchase can be safely used up to the best-before date (usually 35 days from being laid; there will be little change in egg quality during this period), and after this date are recommended for use in baking. Best-before date labelling is compulsory in New Zealand.

Marketing

Marketing involves a range of prices depending on the grade, brand or other differences that attract particular types of consumers: free-range, cage-free, barn, organic, vegetarian-fed and maize-fed are among the choices available. Eggs can also be enriched with selected nutrients by feeding diets containing high concentrations of those nutrients. Such nutritionally enriched eggs (also referred to as designer eggs) are commercially available in many countries, including New Zealand. Among them are eggs with elevated levels of vitamin E, selenium and iodine, eggs

Layer hens outdoors.

with higher contents of polyunsaturated fatty acids, eggs with lower cholesterol levels and eggs containing omega fatty acids.

Production of hatching eggs

Hatching (or fertilised) eggs are produced by parent stock raised in breeder farms. Eggs for the hatching of broiler-type and layer-type chicks are produced in separate breeder farms — broiler breeder farms and layer breeder farms. The production objectives, housing and management of the two breeder types are essentially similar. The objective is to achieve the maximum production of fertile, hatchable eggs. These eggs are not meant for human consumption and are never sold to consumers in New Zealand.

The management of hens for the production of hatching eggs differs from that of table egg production in a number of ways. First, the birds are never caged; a barn housing system, with nesting facilities, is used because hens and roosters have to be reared together to mate and fertilise the eggs. Second, breeder birds are genetically inclined to over-eat and will readily put on weight. An overweight hen will be a poor egg-producer and an overweight rooster will be a low sperm-producer and have difficulty in mating. A restricted feeding programme is therefore typically used to control feed intake to maintain a desired growth rate and body weight. Restriction may be in the form of reduced daily intake, skipped feeding (e.g. skip a day or feeding only five days during a week), or feed quality. Third, the diets offered to breeders will have relatively higher levels of vitamins and trace minerals, which are essential to ensure optimum fertility and hatchability.

Males (cockerels) and females (pullets) are

One-week-old broiler chickens in shed.

reared separately until sexual maturity and are then moved to breeder laying houses. Males and females come from different genetic lines in order to achieve hybrid vigour in the offspring. For the maximum production of fertile eggs, a male to female ratio of at least 1:10 must be maintained. To maintain fertility, young roosters are usually introduced as the flock ages. In breeder hens, egg production starts at 24–26 weeks of age, increases to a peak of 85 per cent at 32 weeks and then gradually declines with age. Hatchability usually peaks (at 90 per cent) between 36 and 40 weeks of age. Each breeder hen will produce 150–180 hatching eggs during a 40-week production period.

High egg quality is critical to ensure high hatchability. Careful barn layout and attention to bird behaviour is required to avoid floor eggs (those laid on the floor rather than in the nesting boxes), and eggs must be collected frequently to avoid soiling. Eggs to be sent to the hatchery are carefully selected, with dirty, undersized and oversized eggs being discarded. The eggs must be stored at a cool temperature (15–18°C) and transferred to the hatchery within a few days.

There are three commercial hatcheries supplying day-old layer chicks in New Zealand: Golden Coast Commercial, operated by Tegel Foods Ltd, supplies Hyline strain; Bromley Park Hatcheries and Heslips Hatcheries supply Shaver strain. Day-old broiler chicks are supplied by Bromley Park Hatcheries (Cobb strain) and Golden Coast Commercial (Ross strain).

Hatcheries incubate fertilised eggs and produce day-old chicks. The hatching of chicken eggs, which takes about 21 days, is a carefully controlled two-stage operation with set conditions for temperature, humidity and ventilation, and regular egg-turning. In the first stage, which lasts 18 days, the eggs are incubated in setters. A modern setter is typically the size of a large room, with a central corridor and racks on either side. The eggs are held relatively tightly (large end up) in trays that are held in the racks. Inside the setter, temperature and humidity are closely maintained. Fans circulate air, as needed, to ensure good ventilation and uniform temperature. The racks tilt from side to side, usually on an hourly basis.

Three days before the eggs are due to hatch (day 18), they are moved into a hatcher unit (stage 2). They are no longer turned, so that the embryos have time to get properly oriented for their exit from the shell, and the temperature and humidity are optimal for hatching. The hatcher is similar to the setter but has larger, flat-bottomed trays so that the eggs can rest on their sides and newly hatched chicks can easily pop out and walk. The typical hatchability rate in a well-managed hatchery is over 90 per cent.

At hatching (days 20–22), the trays are removed from the hatcher. Chicks are inspected, with those that are unhealthy or deformed being disposed of. The chicks are sexed and transported in chick boxes to the farms. Specialised climate-controlled trucks are typically used, depending on climate and transport distance.

Waste management

Litter associated with meat chicken production, manure generated from table egg production (pullets and layers) and dead birds are the three wastes of primary concern in poultry production. One kilogram of litter is produced per broiler in one production cycle, and an average of 16 kg of dry matter is voided by a layer per annum. Based on present poultry production in New Zealand (110 million broilers and 3 million layers), it is thus estimated that approximately 150,000 tonnes of manure wastes are generated annually. Large-scale accumulation of these wastes creates disposal and pollution problems unless they can be used in an environmentally and economically sustainable manner.

Traditionally, poultry wastes are used as an organic fertiliser in vegetable gardens and grasslands to maintain soil fertility. Dried poultry wastes obtained directly as droppings (urine and excreta) from layer units or as a mixture of droppings and bedding material (wood shavings) from broiler units contain considerably higher levels of dry matter than other animal manures and are rich in all nutrients, including micronutrients. Their value as a source of plant nutrients, especially for organic growers, is well recognised.

Another important feature is that the addition of poultry manure also enhances the physical fertility (structure, organic matter content, water-holding capacity and aggregate stability) of soils (Bolan et al., 2000). In New Zealand, the land base available for manure application can be limited, mainly related to the cost of transporting the manure. Consequently, poultry manure is usually applied in the immediate vicinity of poultry farms.

Fresh poultry manure, particularly from cage

systems, is very difficult to handle due to its high moisture content. Its obnoxious odour tends to aggravate the problem by making it unpleasant to handle the waste. Excessive application of fresh manure must be avoided due to its high nitrogen content; the accumulation of ammonia in soils will cause injury to seedlings and roots.

Aerobic composting is most commonly practised to overcome some of the problems associated with the handling and disposal of poultry manure. This reduces the bulkiness of the waste and yields a stabilised product that is suitable for handling and land application. Composting also eliminates animal and human pathogens and could reduce the risks of polluting groundwater.

Commonly used dead-bird disposal practices include burial in pits, incineration and rendering. Co-composting with poultry litter is an option that produces a material amenable to land application.

Layer hens roosting in barn.

Environmental considerations

Emission of greenhouse gases (methane, nitrous oxide and carbon dioxide) and the associated global warming effects are serious public concerns worldwide. With its strong animal production base, New Zealand is unique in its emissions profile. Emissions in New Zealand are dominated by methane and nitrous oxide, in contrast to the rest of the world where carbon dioxide represents over 60 per cent of the emissions (Saggar, 2008). Ruminants (cattle and sheep) are the major contributors to the current emissions.

Exact emission estimates for the local poultry industry are not available, but it is widely accepted that the contribution of poultry is minor compared with other animal production sectors (Ministry of Agriculture and Forestry, 2011). According to a recent study, the emissions and carbon footprint of poultry production, per kilogram of feed consumed, are roughly half those of pork, one-quarter those of beef and nearly a seventh those of lamb (Ministry for the Environment, 2015). These reports, taken together, suggest that poultry account for less than 1 per cent of New Zealand agricultural emissions.

References

Ambler, G.H. (1923). *Utility poultry farming in New Zealand.* Christchurch: Whitcombe & Tombs Ltd.

Baker, M.G., Sneyd, E., & Wilson, N.A. (2006a). Regulation of chicken contamination urgently needed to control New Zealand's serious campylobacteriosis epidemic. *New Zealand Medical Journal*, 119, 1–8.

Baker, M.G., Wilson, N.A., Ikram, R., Chambers, S., Shoemack, P., & Cook, G. (2006b). Is the major increase in notified campylobacteriosis in New Zealand real? *Epidemiology and Infection*, 133, 1–8.

Binney, B.M., Biggs, P.J., Carter, P.E., Holland, B.M., & French, N.P. (2014). Quantification of historical livestock importation into New Zealand. *New Zealand Veterinary Journal*, 62, 309–14.

Bolan, N., Mahimarajah, S., & Ravindran, V. (2000). Potential value of poultry manure as a nutrient source and soil amendment. In V. Ravindran (Ed.), *Proceedings of the New Zealand Poultry Industry Conference, Palmerston North*, vol. 5 (pp. 61–65). Palmerston North: World's Poultry Science Association, New Zealand Branch, and Monogastric Research Centre, Massey University.

Brooks, M. (2016). New Zealand poultry industry: issues and legislation updates. In M.R. Abdollahi and V. Ravindran (Eds), *Proceedings of the Advancing Poultry Production — Massey Technical Update Conference*, vol. 18. Palmerston North: Monogastric Research Centre, Massey University.

Brooks, M. (n.d.). Poultry Industry Association of New Zealand, Auckland (mimeo).

Cobb. (2014). *Broiler performance: a nutrition guide.* Siloam Springs, Arkansas: Cobb-Vantress. Accessed at www.cobb-vantress.com/docs/default-source/cobb-500-guides/Cobb500_Broiler_Performance_And_Nutrition_Supplement.pdf.

Cooper-Blanks, B. (1999). *The New Zealand poultry meat industry: an education and industry resource.* Auckland: Poultry Industry Association of New Zealand.

Cundy, M. (2010). WPSA activities in New Zealand — an historical view. In V. Ravindran (Ed.), *Proceedings of the New Zealand Poultry Industry Conference, Palmerston North*, vol. 10 (pp. 10–18). Palmerston North: World's Poultry Science Association, New Zealand Branch, and Monogastric Research Centre, Massey University.

Diprose, R. (2001). New Zealand's poultry has a superior health status. *World Poultry*, 17(6), 12–14.

Duncan, G.E. (2014). Determining the health benefits of poultry industry compliance measures: the case of campylobacteriosis regulation in New Zealand. *New Zealand Medical Journal*, 127, 22–37.

Gordon, F.E.A. (1908). *Utility poultry farming for Australasia — a book for the man with 6 birds or 6,000.* Christchurch: Whitcombe & Tombs Ltd.

Hall, N., & Clarke, S. (Eds). (2011). How to care for your poultry. Auckland: Fairfax New Zealand.

Hyline. (2016). *W–36 commercial layers — management guide.* West Des Moines, Iowa: Hyline. Accessed at www.hyline.com/UserDocs/Pages/36_COM_ENG.pdf.

Kissling, J.H. (1945). The poultry industry in New Zealand. *New Zealand Poultry World*, 66–67.

Ministry for the Environment. (2015). New Zealand's greenhouse gas inventory 1990–2013. Report submitted to the United Nations Framework Convention on Climate Change. Wellington: Ministry for the Environment.

Ministry of Agriculture and Forestry. (2011). Emission estimations for the commercial chicken, non-chicken and layer industries within New Zealand. MAF Technical Paper no. 2011/22. Wellington: Ministry of Agriculture and Forestry.

National Animal Welfare Advisory Council (NAWAC). (2010). Animal Welfare (Commercial Slaughter) Code of Welfare 2010. Wellington: Ministry for Primary Industries.

National Animal Welfare Advisory Council (NAWAC). (2011). Animal Welfare (Transport within New Zealand) Code of Welfare 2011. Wellington: Ministry for Primary Industries.

National Animal Welfare Advisory Council (NAWAC). (2012a). Animal Welfare (Layer Hens) Code of Welfare 2012. Wellington: Ministry for Primary Industries.

National Animal Welfare Advisory Council (NAWAC). (2012b). Animal Welfare (Meat Chickens) Code of Welfare 2012. Wellington: Ministry for Primary Industries.

Ravindran, G. (2012). Egg: Nature's unique miracle food. Recent developments and future possibilities. In V. Ravindran (Ed.), *Proceedings of the New Zealand Poultry Industry Conference, Hamilton*, vol. 11 (pp. 146–58). Palmerston North: World's Poultry Science Association, New Zealand Branch, and Monogastric Research Centre, Massey University.

Ravindran, V. (2013). Poultry feed availability and nutrition in developing countries. Poultry Development Review. Rome, Italy: Food and Agriculture Organisation of the United Nations. Accessed at www.fao.org/ag/againfo/themes/en/poultry/ AP_nutrition.html.

Ross. (2014). Ross 308 broiler nutrient specifications. Midlothian, Scotland: Aviagen. Accessed at en.aviagen.com/assets/Tech_Center/Ross_Broiler/Ross308BroilerNutritionSpecs2014-EN.pdf.

Saggar, S. (2008). Global climate change, greenhouse gas emissions and contributions from poultry production in New Zealand. In V. Ravindran, *Proceedings of the New Zealand Poultry Industry Conference, Palmerston North*, vol. 9 (pp. 169–84). Palmerston North: World's Poultry Science Association, New Zealand Branch, and Monogastric Research Centre, Massey University.

Sears, A., Baker, M.G., Wilson, N., Marshall, J., Muellner, P., Campbell, D.M., Lake, R.J., & French, N.P. (2011). Marked campylobacteriosis decline after interventions aimed at poultry, New Zealand. *Emerging Infectious Diseases* (online), 7(6). Accessed at http://dx.doi.org/10.3201/eid1706.101272.

Shaver. (2015). *Shaver commercial management guide.* Peterborough, UK: Hendrix Genetics. Accessed at arya-dan-roshd.ir/en/images/general/articale/201111%20Shaver%20White%20commercial%20management%20guide%20North%20American%20version%201.pdf.

Stone, W.R. (1972). The poultry industry of New Zealand. *New Zealand Veterinary Journal*, 20, 142–45.

Thomson, G.M. (1922). *The naturalization of animals and plants in New Zealand.* London, UK: Cambridge University Press.

Wintle, V., & Lepper, S. (2012). Poultry industry. Te Ara Encyclopaedia of New Zealand. Accessed at www.TeAra.govt.nz/en/poultry-industry.

Chapter 8

Horse Production

Chris Rogers, Erica Gee and Charlotte Bolwell

Chapter 8

Horse production

Chris Rogers, Erica Gee and Charlotte Bolwell
Institute of Veterinary, Animal and Biomedical Sciences
Massey University, Palmerston North

Introduction

The horse is an important domestic animal in New Zealand and is used in many ways. Although horses are valuable production animals, what they produce is of a different nature to other livestock. Horses produce value through racing or sport or through the breeding of race or sport horses. A small number are used as work horses on high-country stations and a very few are used as draught horses. The list of sports in which horses are used is a long one, and many of the animals bred in New Zealand are exported because of their internationally recognised value.

Horses have an interesting position on the human–animal inter-relationship scale, being considered livestock by some and a type of pet by others. Horses that become unable to work because of injury or old age are usually slaughtered, either for human consumption or as pet food. The number of horses processed for human consumption fluctuates between years; approximately 1500 horses are processed annually, with all meat going to the export market. However, increasing numbers of animals are being kept until they reach old age and are then euthanased as pets rather than as a source of meat.

The geography and the social structure of New Zealand, especially in provincial and rural parts of the country, make horse ownership easy; neither exclusive nor terribly expensive (Figure 8.1). On a per capita basis, New Zealand has a high horse population, with an estimated 30 horses per 1000 people and 3.7 horses per 1000 hectares (ha). In comparison, Ireland — also a horse-friendly country — has fewer horses per head of population (19 horses/1000 people) but more horses per unit of land (11.4 horses/1000 ha) (Liljenstolpe, 2009).

The use of horses can be divided into five major categories: Thoroughbred racing, Standardbred (harness) racing, sport, leisure/recreation and work (Table 8.1). A number of feral and semi-feral populations of horses also exist, the most recognisable being the Kaimanawa horses.

The two forms of racing are tightly controlled, and both Thoroughbred and Standardbred horse management and racing have been investigated scientifically. There is less information available about sport horses (identified as horses bred or used within the Olympic equestrian sports of dressage, show jumping and eventing), and even less on leisure and work horses. This chapter will differentiate between these five uses of horses wherever possible.

Table 8.1 **The uses of horses and the numbers involved within New Zealand**

Use	Number
Thoroughbred industry	
Broodmares	8084
Sires	250
Annual foal crop	4500
Young stock not yet registered for racing	4500
Racehorses	5576
Total Thoroughbred industry racing and breeding	*22,910*
Standardbred (harness racing)	
Broodmares	4000
Sires	200
Annual foal crop	2800
Young stock not yet registered for racing	2800
Racehorses	3000
Total Standardbred industry racing and breeding	*12,800*
Sport horse industry	
Broodmares	4000
Sires	300
Annual foal crop	1000
Sport horses young stock	6000
Total sport horse industry competition and breeding	*18,443*
Other equestrianism and leisure/recreation	
Unregistered horses/ponies	21,429
Quarter Horse/Western	3000
Miniature	3000
Stationbreds (utility/farm horses)	3000
Trekking/recreation	14,286
Total population	*98,868*

Adapted from Rogers & Vallance (2009).

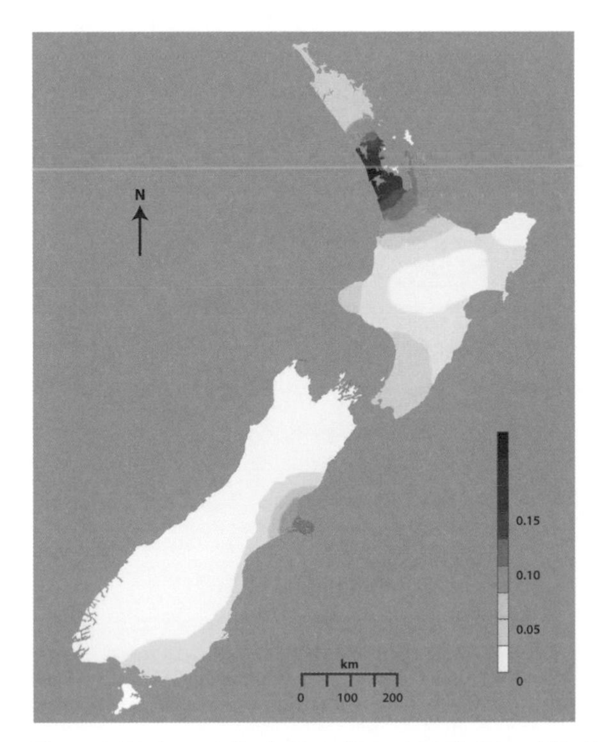

Figure 8.1 Estimated distribution of horse populations within New Zealand. The shading represents the density of horses per 10 km^2. Source: adapted from Rosanowski (2012).

A brief history of the horse industry

The horse has played a major role in the development of New Zealand since the 1840s. Prior to mechanisation, the most common type of horse was a utility horse — a crossbred horse, often from a draught mare bred to a lighter stallion such as a Thoroughbred or an Arab. Draught horses, usually Clydesdales, traditionally provided the horsepower required for agricultural and haulage work.

The advent of cars, trucks and mechanisation on farms, especially after the 1940s and 1950s, led to a dramatic decrease in the horse population (Mincham, 2011). This was accompanied by a significant change in the structure of the horse population in relation to breed and type. Draught horses became rare; so much so that the once common Clydesdale breed is now classified as rare and endangered. On the steeper hill-country sheep stations, the utility horse (now commonly called a stationbred) was still used for stock work, but with the arrival of motorbikes and all-terrain vehicles (ATVs) in the 1980s it also declined in number. More recently, however, a number of hill-country sheep stations have returned to using horses for general mustering and stock work due to their advantages over motorbikes and ATVs on steep country.

In the 1950s, equestrian sports were organised on a national level following the formation of the New Zealand Pony Club and the New Zealand Horse Society. This resulted in an increase in the number of horses kept solely for sport and recreation. In the early 1980s, the importation of warmblood (particular types and breeds of middle-weight horse suited to equestrian disciplines) or sport horse sires from Europe saw the emergence of a semi-commercial industry focusing on the breeding of horses for show jumping, dressage and eventing (Rogers & Wickham, 1993). Today, in a pattern similar to that in other countries, the majority of the horses in New Zealand (approximately 100,000) are recreational or racing horses, and only a small percentage (approximately 3 per cent) are working horses (Table 8.1) (Rosanowski et al., 2012).

Horse racing, especially Thoroughbred racing, has a long and established history in New Zealand. The first Thoroughbred stallion is reported as being imported in 1840. The first official race meeting was in Wellington that same year, and the subsequent rapid expansion of racing clubs throughout the country resulted in the formation of the New Zealand Racing Conference in 1887. This was the predecessor to New Zealand Thoroughbred Racing Inc., the official body that regulates Thoroughbred racing and breeding in New Zealand today.

At one stage, New Zealand had more racing clubs and race tracks per head of population than any other country (Costello & Finnegan, 1988). However, rationalisation of the industry since the 1980s has seen the centralisation of tracks and the merging of racing clubs, generally around the larger population centres (Bolwell et al., 2014a).

The New Zealand horse industry

In many countries, the production of horses is primarily for the domestic market. In contrast, the New Zealand equine industry has a major export focus, with the Thoroughbred industry having the greatest emphasis on international markets. It is estimated that approximately 40 per cent (approximately 1800 out of 4500) of the annual Thoroughbred foal crop is exported, either as yearlings or as two-year-old racing prospects. Each year, about 1400 yearlings are offered for sale at the national sale series, with many of these being destined for the export market. The majority of the Thoroughbreds are exported to Australia (65 per cent); a large proportion of the remaining exports go to Hong Kong, Singapore and Macau (Waldron et al., 2011).

The harness racing industry (Standardbreds) has a more domestic-market focus than the Thoroughbred industry. A large proportion of the harness racing participants breed horses, which they will train and race themselves (Bolwell et al., 2014b). Yearling sales for Standardbreds are becoming more popular, although the proportion of the foal crop sold as yearlings (approximately 500) is low relative to the annual foal crop (approximately 2000 foals born/year). The Standardbreds that are exported (about 15 per cent of the annual foal crop) tend to be older, established racehorses for the Australian and United States racing industries.

Information on the contribution of the sport horse sector to the economy is less precise than that available for the racing industries. It is difficult to estimate the number of horses exported

Table 8.2 **Relative size and economic contribution of the three main sectors of the New Zealand equine industry**

	Harness/Standardbred	Thoroughbred	Equestrian sport
Economics			
GDP ($NZ)	0.25 billion	0.85 billion	1.2 billion
Major export markets	Australia	Australia, Hong Kong, Macau, Singapore	USA, UK, Asia
Number of horses exported (% foal crop)	300 (15%)	~2000 (40%)	100–200 (4%)
Organisation/participation			
National governing body — racing and sport	Harness Racing New Zealand	New Zealand Thoroughbred Racing	Equestrian Sports New Zealand
National governing body — breeding	Harness Racing New Zealand	New Zealand Thoroughbred Racing	Multiple registration societies
Full-time equivalent people involved	16,757	30,554	N/A
Trainers (jockeys/drivers)	1000 (800)	800 (140)	~4000
Horse participants, racing or registered	3600	5800	~5000
Horse breeding			
Number of sires	97	94	300
Broodmares	3700	5800	~6000
Annual foal crop	2000	4000	~4000

North Island pony club show-jumping championships. **Previous:** Broodmares at pasture.

annually. Most of these are show jumpers and eventers. The most lucrative markets for sport horses are the United States and Europe, although the majority of exports are to Asia.

It is estimated that the racing (Thoroughbred and Standardbred) and sport horse industries generate over NZ$2 billion in gross domestic product (approximately 2 per cent of New Zealand's total GDP). A large portion of the economic contribution from racing is driven by exports and by participation, with associated gambling, in horse racing. In the sport horse sector, the financial return per horse is lower than with racing, and the prize money available is likewise lower. Despite the limited opportunities to generate a return on investment, there is a disproportionate level of expenditure within this sector of the equine industry. A significant domestic industry also focuses on providing equipment, clothing and packaged horse feed for the sport horse sector.

Organisation of the horse industry

In New Zealand, the Racing Board provides over-arching governance for racing (Thoroughbreds, Standardbreds and Greyhounds) and management of the Totalisator Agency Board (TAB), the body which controls legal gambling on racing and sport. The New Zealand Racing Board reports directly to the Minister for Racing via the Department of Internal Affairs.

The day-to-day management and regulation of Thoroughbred racing and breeding is controlled by New Zealand Thoroughbred Racing Inc. Harness racing and the breeding of Standardbreds is controlled by Harness Racing New Zealand. Due to the close relationship between New Zealand and Australian racing, and the frequent exchange of horses between these countries, both New Zealand racing codes are in close alignment to the rules for racing and breeding in Australia.

Internationally, for Thoroughbreds, racing is governed by the International Federation of Racing Authorities, which provides guidelines on best practice for the structure and regulation of racing. The breeding of Thoroughbreds is controlled by the International Studbook Committee, which provides regulation on breeding and prohibits the use of artificial insemination (AI) and embryo transfer (ET).

There is no limit on the use of AI within harness racing, and up to one embryo per year may be collected from a mare. Internationally, the guidelines for the regulation of harness racing are based on those of the United States Trotting Association. In a similar way to the Thoroughbred industry, there are major similarities between the harness racing regulations in Australia and New Zealand.

Equestrian sport in New Zealand is controlled by Equestrian Sports New Zealand (ESNZ), which is responsible for show jumping, dressage, eventing, para-equestrian and endurance racing. The New Zealand Pony Club Association is affiliated with ESNZ. The rules for equestrian sport are based on those of the international governing body, the International Equestrian Federation (FEI). In New Zealand the breeding of sport horses is unregulated and is split between various breeding societies and organisations (Creagh et al., 2012). There are no restrictions on the use of assisted reproductive technologies and many horses are the product of imported frozen semen (George et al., 2013).

Horses used for leisure and recreation are believed to constitute the majority of the horses in New Zealand. These horses are not required to be registered with any governing body. Many compete in what are unofficial events at a lower level than that offered by ESNZ-sanctioned or official shows. The large number of leisure horses and riders (approximately 70 per cent of horses and riders) is typical of most equestrian populations.

A small number of horses are used on stations for stock work. They are not registered and are bred according to need. The breeding of these horses is still based on the interbreeding of a variety of breeds to produce a smallish, docile stock horse. Traditionally, the genetics in this type of horse were based on Clydesdale, Thoroughbred and Arab horses. More recently there has also been use of Standardbred, warmblood, Quarter Horse and Appaloosa genetics.

Many of the smaller breed societies are affiliated with the Royal Agricultural Society; these include such diverse breeds as Clydesdales, riding ponies, gypsy cobs and miniature horses. The proliferation of smaller specialist horse or pony studbooks is an international trend observed in many countries and creates logistical difficulties in maintaining critical mass for a sound genetic base and from a management perspective (Aurich & Aurich, 2006). Within New Zealand, many of the small niche studbooks or breed groups struggle to maintain an efficient or structured pedigree database and management system (Simmons, 2015).

Recreational riding.

Pasture-based production system

The temperate climate in New Zealand permits pasture-based production systems for horse breeding and management. This unique feature gives rise to a number of opportunities to grow and manage horses in a system that is close to the ecological niche from which the horse evolved. A pasture-based management system reduces production costs and reliance on concentrate feed to promote growth in young stock (Rogers et al., 2007).

The pasture on most stud farms and commercial equine properties is a perennial ryegrass/white clover (*Lolium perenne/Trifolium repens*) mix (a ratio of 85:15), which horses find highly palatable (Hirst, 2011; Randall et al., 2014). The dry matter digestibility of fresh pasture based on perennial ryegrass is typically 60–65 per cent (Grace et al., 2002a, 2002b, 2003). The digestible energy (DE) and protein content of this pasture is high compared with North American pastures: in summer it typically provides 10.8 megajoules (MJ) of digestible energy per kilogram of dry matter (kg DM) and 18.6 per cent crude protein (Hirst, 2011). The period of the most rapid pasture growth (providing approximately 60 kg DM/ha/day) coincides with the first weeks following birthing for most commercial Thoroughbred horses (October to November), and it remains at about 30 kg DM/ha/day through to when the foals are weaned (April).

The macro-element composition of typical New Zealand pastures meets the nutritional requirements of lactating broodmares and weanling foals (Grace et al., 2002a, 2002b, 2003). However, the concentration of copper in New Zealand pasture is frequently below the National Research Council (NRC) minimum requirements of 10 mg/kg DM (National Research Council, 2007). Oral copper supplementation of mares in late pregnancy has been associated with less evidence of bone and cartilage lesions in Thoroughbred foals at five months of age (Pearce et al., 1998a). However, supplementation of Thoroughbred foals with copper had no significant effect on bone and cartilage lesions in foals at five months of age compared with unsupplemented foals (Pearce et al., 1998b).

The pasture concentration of calcium may also be marginal according to NRC recommendations (National Research Council, 2007), but does not limit the bone development or growth of Thoroughbred foals (Grace et al., 2003); this may be due to the calcium in pasture having high bioavailability (Hoskin & Gee, 2004).

Diet for Thoroughbred broodmares and young stock

Broodmares are managed year round at pasture and in paddocks with a perennial ryegrass/clover sward. The primary nutritional source for broodmares is pasture, supplemented with hay during the winter months. The typical requirements of the broodmare are estimated to be approximately 76 megajoules of digestible energy per day (MJ DE/day). Broodmares on commercial farms are provided with pasture having an average sward height of 10 cm (equivalent to approximately 3500 kg/DM/ha) (Rogers et al., 2007). In late pregnancy the requirements of the broodmare increase approximately 10 per cent, which can be met by the pasture on offer. During the third trimester of pregnancy, most farms will provide mineral pellets; these generally have low energy (DE) but provide additional minerals, especially copper, zinc, calcium and phosphorus.

The DE requirements of a weanling are estimated to be approximately 60 MJ/day. Despite being managed on pasture with a typical sward height of 10 cm (i.e. approximately 3500 kg/DM/ha), weanlings are often provided with concentrate feed (about 3 kg [range 1–6 kg] per day as fed) (Stowers et al., 2009). Typically this equates to 47 MJ DE from concentrate feed, despite the existence of studies demonstrating equivalent growth rates from a pasture-only diet.

Diet for Thoroughbred racehorses

Racehorses are generally managed in an intensive production system, similar to other international racing jurisdictions, with stabling and some turnout in small turnout yards or paddocks (Williamson et al., 2007). This is in part due to space constraints and in part due to the need to keep the management of the horse consistent, with limited changes in diet and management as it travels nationally and internationally for racing.

In New Zealand, most racehorses are fed commercial pre-mixed diets rather than straight grains (oats only). The usual quantities fed per day are 5.5 kg of pre-mix and 2.5–4.5 kg of hay (Williamson et al., 2007). These values are similar to those reported for Australia and the United States, reflecting the similarities in racehorse management across racing jurisdictions.

Diet for sport horses

Sport horses in New Zealand are also managed at pasture, or with a combination of pasture and stabling (Verhaar et al., 2014). The management of sport horses in groups at pasture is in stark contrast to that of similar horses in Europe, where stabling is the common management technique and limited turnout is provided (a few horses are turned out for free exercise in the company of other horses).

When at pasture, New Zealand sport horses are reported to have pasture cover in the range of 1800–2000 kg DM/ha and are provided with approximately half of their DE requirements (about 60 of 116 MJ/day) as concentrate and supplementary forages (e.g. hay). Similar to racehorses, the majority of sport horses are fed commercial pre-mix concentrate feed. Based on the estimated quantity of the pre-mixed feeds offered, sport horses would consume approximately 3.52 kg DM/day in pasture (Verhaar et al., 2014).

In contrast to racehorses, leisure horses and pony club ponies are often kept at pasture virtually all the time (Fernandes et al., 2014). Many owners of ponies restrict pasture access in an attempt to reduce feed intake for what may be efficient feed converters (Fernandes et al., 2015). Despite many of these horses having a low DE requirement due to being over-fat, and having a low exercise level, many owners still also provide a small quantity of concentrate feed (Fernandes et al., 2014).

Growth and development

Most of the data on the growth of horses in New Zealand are for the Thoroughbred. At birth, colt (male) foals are typically heavier than filly (female) foals (56 kg vs 54 kg). The period of the most rapid growth is prior to weaning (4–6 months old), with growth rates of up to 2 kg/day during the first few weeks following birth and then 1 kg/day at the time of weaning (Figure 8.2) (Morel et al., 2007). After weaning, the growth rate drops to 0.6–0.7 kg/day, and by the time the Thoroughbred is presented for sale at the yearling sales (14–15 months old) it will be 90–95 per cent of its mature height and about 80 per cent of its mature weight.

The ability to manage horses at pasture all year round is believed to promote stimulation of the musculoskeletal system and this, combined with the ability to provide the bulk of the required nutrition from pasture, is believed to reduce the risk of osteochondrosis developing (a failure of the bone underlying the smooth cartilage of the joints) (Van der Heyden et al., 2013).

A cross-sectional survey of commercial stud farms in Australia and New Zealand indicated that the radiographic prevalence of osteochondrosis in New Zealand Thoroughbred yearlings was 13 per cent, half that estimated for a similar group of animals in Australia (Castle, 2012). The ability to have foals freely exercise at pasture (prior to weaning) when the musculoskeletal system is most receptive to stimuli promotes the positive development of cartilage and chondrocyte viability (Dykgraaf et al., 2008). The advantages of early exercise and pasture management have been reported in a number of reviews (Rogers et al., 2012, 2014a).

There is limited data on the growth and development of Standardbreds in New Zealand. The data that does exist indicates lower birth weights and slower growth rates, in part due to the more conservative management of Standardbred young stock compared with

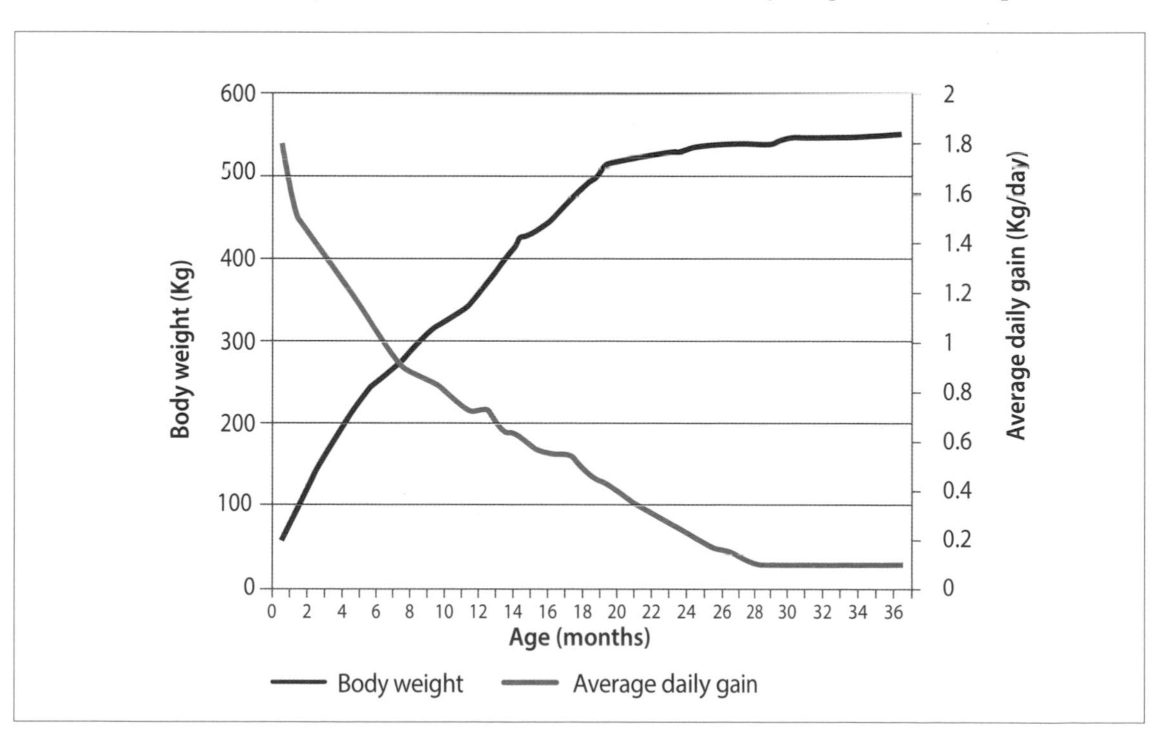

Figure 8.2 Mean body weight (kg) and average daily gain (kg/day) from a cohort of New Zealand Thoroughbred foals. Source: Rogers et al. (2008).

Thoroughbreds (Stowers et al., 2010). There is no data on sport horse growth and development in New Zealand, although overseas data indicates that these horses have higher birth weights and greater mature weights and heights than both the Standardbred and the Thoroughbred (Rogers et al., 2014a).

Wastage — the losses associated with morbidity (incidence of disease) and mortality during growth and development — appears to be greatest before weaning; about 10 per cent of all Thoroughbred foals born are not officially registered as weanlings (Rogers et al., 2009, 2016). Data from Ireland and the United States indicates that the period of greatest risk during pre-weaning is the first 15 days following birth, with mortality rates of 5 per cent and morbidity rates of 25 per cent (Cohen, 1994; Morley & Townsend, 1997).

Thoroughbred yearling preparation and sales

An intensive period of management for growing Thoroughbreds is associated with the preparation of yearlings for the annual yearling auctions in February. Most farms will start this preparation the preceding November. Yearlings will spend about 12 hours in stables (loose boxes) and 12 hours at pasture; in addition there will be a focus on feeding and handling to improve the physical condition of the horses so that they are well grown and demonstrate peak condition for the sales (Bolwell et al., 2010a). The length and intensity of yearling preparation differs depending on the expected sales price of the yearling; the most intensive management being for yearlings intended for the Premier (elite) sales series.

Racing and sport

Thoroughbred racing

The majority of the prize money in Thoroughbred racing is focused around the flat races specifically for two- and three-year-old horses. The three-year-old season is often called the classic season, as it is during this year that horses can race in the five most prestigious races (the classic races) — the Derby, the Oaks, the 1000 Guineas and 2000 Guineas, and the St. Leger. These races belong to the top 5 per cent of races run in New Zealand and are classified as group or listed races.

Horses that win or are placed in group or listed races have their name listed in bold type in pedigree catalogues and are referred to as having 'black type'. It is important for a horse to achieve black type status if it is to subsequently be a commercial breeding prospect. The quantity of black type within a mare's pedigree is one of the criteria used to decide whether her progeny will be sold in elite or lower-quality yearling sales.

In New Zealand, most Thoroughbred horses enter race training as two-year-olds and the majority start racing as three-year-olds. Many horses start being educated (breaking-in and pre-training) soon after the yearling sales series, by trainers specialising in the education of young horses. After a short spell (turnout time at pasture), they start another couple of weeks of pre-training before going to the racing trainer.

Mare at pasture.

Most trainers take 10–12 weeks from the two-year-old entering the racing stable until its first trial or race start; the pattern of training is similar between most trainers (Bolwell et al., 2010b).

Although most racehorses will enter training as a two-year-old, only about 30 per cent of them will have a race start in this age class. This is in part due to most trainers having a primary focus on education and conditioning, rather than racing their horses as two-year-olds, and an alternative focus of producing a two-year-old race trials winner (a race trial is an unofficial qualifying type of race) that can be sold into Asia (Bolwell et al., 2010b, 2012). An early start to education and conditioning may also be due to recognition of the positive association of early education and conditioning on career length and success in racing (Tanner et al., 2011, 2013).

The majority of Thoroughbred horses have a race start as a three-year-old in New Zealand, with many only having a few starts in their life (8–11 starts). The racing population is slightly skewed towards the older horses (4–5 years old), due to the weight for age (WFA) handicapping system that permits horses of different ages and ability to compete on an equal footing in the same race.

Approximately 2900 flat races and 120 jumps races are run each year, with jumps racing predominantly restricted to the winter season. The median race distance for flat races is 1400 metres, while jumps races are longer with a median distance of 3100 metres (Bolwell et al., 2014a). All racing is conducted on turf tracks, the majority of which are oval in shape, approximately 1800 metres in circumference and raced counter-clockwise (Rogers et al., 2014b).

Standardbred racing

The process of training and educating Standardbreds for harness racing is similar to that in the Thoroughbred industry, with the exception that almost half of the horses in training will be with an owner/trainer. The owner/trainer system means that there is greater participation of amateur trainers than is seen in the Thoroughbred industry. The owner/trainer nature of harness racing is believed to provide greater variation in the structure of the training pattern for harness-racing horses, together with a skewing of the racing population to older horses with a greater number of starts in life (19–43 starts) than is seen in Thoroughbred racing (Bolwell et al., 2014a, 2014b).

Wastage of Thoroughbred and Standardbred horses

Historically there has been significant supply chain wastage within the racing industries, with 30 per cent of the Thoroughbred foals born not having a race start. Since the global economic crisis of 2007–08 there has been a reduction in the number of Thoroughbred foals born but not in the number of horses racing, and thus the supply chain wastage has reduced (Rogers et al., 2016). This change has not, however, altered the proportion of horses lost from the racing industry after starting racing. Approximately one-third of Thoroughbred horses entering race training will be retired for voluntary reasons (generally lack of talent), and a further third will be lost due to involuntary reasons (78 per cent of these are due to musculoskeletal injury and lameness) (Perkins et al., 2005a).

In relation to lost training days, two-year-old Thoroughbreds have the most days lost, generally due to shin soreness (dorsal metacarpal disease). The incidence of shin soreness is related to the rate of increase in workload, especially the delayed introduction of high-speed work. While two-year-olds have the most training days lost, it is the older horses (five years and older) that have the greatest risk of injury, in particular injuries related to the accumulation of workload cycles (bone fracture and superficial digital flexor tendon injury).

After their racing career has ended, the better-performing horses (mares and some colts) enter the breeding herd. Many Thoroughbred geldings (castrated males; about 45 per cent) move into equestrian sport (Verhaar et al., 2014). In contrast to Europe, the Thoroughbred makes up a large proportion of the horses used for equestrian sport in New Zealand, particularly eventing (approximately 89 per cent of horses), and many horses used in sport are bred from Thoroughbred mares (Rogers & Firth, 2005). Of those not suitable for higher-level sport, many are used for recreational sport and others are processed for pet food or for human consumption.

The wastage and production loss figures within the harness racing industry are similar to those reported for Thoroughbreds (Tanner et al., 2012). In one study that followed an entire foal crop, 33 per cent of the foals born were not registered with a trainer and only 44 per cent had a race start by the end of the three-year-old racing season (Tanner et al., 2011). There are limited post-racing careers for harness-racing horses: some are used in trekking enterprises or as farm hacks, but the majority are processed for meat.

Equestrian sport

Based on the number of horses and riders registered, the dominant equestrian sport in New Zealand is show jumping (approximately 1800 horses), followed by dressage (approximately 1100 horses) and then eventing (approximately 850 horses). Internationally, most equestrian sport is focused around the disciplines of show jumping and dressage. The large number of participants (and proportion of horses) in eventing is unique to New Zealand and may be due to cultural bias, the abundance of Thoroughbreds that are well suited to the sport, and the availability of open farmland on which eventing competitions can be held. In contrast to Europe, most sport horses in New Zealand are geldings (approximately 70 per cent) rather than the more typical 50:50 ratio of mares and geldings seen elsewhere. The majority of sport horse exports from New Zealand are show jumpers and event horses.

The genetic basis of the horses used for dressage tends to be very homogenous. Most dressage horses are of warmblood breeding, from either domestically based stallions or imported frozen semen. Many riders competing at the upper level of the sport have imported warmblood horses from Europe (Friedrich et al., 2011). Because of the structure of the sport and the specialised breeding programme, most dressage horses start competition and training as three-year-olds.

The majority of event horses are Thoroughbreds (89 per cent) and have entered eventing after a previous career as a racehorse. The minimum age for starting eventing is five years; many horses start the sport after finishing their racing career as three- or four-year-olds and then having a period of retraining to be suitable as event horses.

Show-jumping horses in New Zealand are genetically and phenotypically the most heterogeneous population. Most are warmbloods or sport horses, although a large proportion of

Thoroughbred and stationbred horses are also used for jumping. The average age at entry to the sport is five years even though the earliest jumping competitions are for four-year-olds, often indicating that a number of horses enter show jumping after being started at another sport.

The median career length of sport horses is similar to that found in Europe, being between three and four years (Friedrich et al., 2011). Within New Zealand there is limited data on injury and reasons for wastage in sport horses, although data from Europe implies that, similar to racing, the major reason for loss is voluntary retirements (lack of talent), and musculoskeletal injury is the major reason for involuntary losses (Dijkstra et al., 2016).

Ponies and pony clubs

Most New Zealand riders learn to ride as children on ponies. Ponies are described as being less than 14.2 hands (about 1.48 metres) high at the withers (the highest point on a horse's back, at the base of the neck, above the shoulders). The majority of ponies are often described as mixed breed, as most breeding is aimed at producing a general-purpose pony rather than with the intention to maintain purity of genetics or breed preservation (Fernandes et al., 2014). This heterogeneous pony population generally consists of mixes of the English pony breeds, Connemaras and small stationbred horses. The resulting genetic mix results in a thrifty type of horse that is an efficient feed convertor; this may explain the higher prevalence of obesity observed in this population (Fernandes et al., 2015).

The organisation responsible for junior rider (child) sport and education is the New Zealand Pony Club Association (NZPCA), which has approximately 15,000 registered members. It is the largest of the equestrian sport organisations in New Zealand. The NZPCA is affiliated with ESNZ, the body responsible for organised equestrian sport and for reporting to the Olympic and high-performance sport committees.

Breeding

The breeding of racing horses

Within the racing codes (Thoroughbred and Standardbred) there has been significant consolidation in the breeding herd, with a reduction in the number of Thoroughbred sires at stud over the past 25 years from 265 to 94. Associated with this decrease have been an increase in the reproductive output of these sires (the average book size has increased) and a greater proportion of the sires covering (mating with) 100 or more mares in a breeding season (September to December).

There has also been a reduction in the national Thoroughbred broodmare herd, but (as observed with the sires) there has been a corresponding increase in the reproductive capability of the mares (10,176 mares and 5882 foals vs 5826 mares and 3927 foals, respectively, in the 1989/90 and 2011/12 seasons). The reproductive efficiency of the mares has increased from 57 foals registered per 100 mares mated to 67 foals registered per 100 mares mated (Rogers et al., 2009, 2016).

The primary driver of the increased female

reproductive efficiency has been a greater emphasis on selection for younger broodmares (less than 12 years old) (Rogers & Gee, 2011). There has also been an associated increased commercial focus on the breeding of mares that have a 'commercial pedigree' and will appeal to buyers of the resultant yearling. The main selection focus to achieve this has been to use only those broodmares that have themselves been a race winner.

Within the Thoroughbred industry, increases in reproductive output must be obtained without the use of AI or ET. These limitations mean that a popular Thoroughbred stallion, which is covering more than 100 mares in a breeding season, may only have the opportunity to serve a mare once in a given breeding season. Due to the management constraints of breeding large numbers of mares without AI, a mare that is bred late in the breeding season and fails to conceive may not have a chance to be bred with that stallion in that breeding season. To maximise the chances of conception, mares that breed with a popular commercial sire tend to be younger and are selected for reproductive health.

In the Standardbred industry, both AI and ET are permitted and this reduces the physiological challenge for the stallion during the breeding season. Semen may be collected from a popular harness-racing breeding sire 3–4 times a week, whereas a Thoroughbred sire may cover 3–4 mares per day at the height of the breeding season. A collection from a Standardbred stallion may be used to inseminate from seven to 20 mares.

Breeding calendar for Thoroughbred horses

The equestrian calendar starts on 1 August, and in New Zealand this is the official birth date of all horses (Figure 8.3). The objective of most commercial racehorse breeders is to try to have a foal born as close to this date as possible. On commercial breeding farms during August, empty mares (those that are not pregnant) will be managed to promote the transition from anoestrus to oestrus. Generally, empty mares are managed at a lower body condition score than pregnant mares, and during August the empty mares are provided with an increasing plane of nutrition and will be rugged (given horse covers). The transition to oestrus may also be promoted with the use of hormones and/or light therapy.

Horses are long-day breeders (breeding naturally when the number of daylight hours is higher), so exposure to increased light (artificial lighting either in yards or stables) or the use of special blue-light transponders will promote the seasonal transition from anoestrus to oestrus earlier in the season. Most commercial farms actively try to have empty mares cycling (undergoing normal oestrus cycles) early in the season so that they can actively manage the number of mares a stallion will need to cover in the second half of the breeding season when more recently foaled mares are bred (Rogers et al., 2007).

The gestation period of horses in New Zealand is typically longer than for horses in the northern hemisphere (approximately 349 days). Mares bred earlier in the season have longer gestation lengths than those bred later in the season, possibly so that their foals are more likely to be born when pasture growth is higher during October/November (Dicken et al., 2012). To maintain a yearly foaling pattern, mares need to become pregnant within 25 days of foaling; otherwise, foaling dates will drift to become too late in the season. To achieve this, some mares will be short-cycled (returned to oestrus using

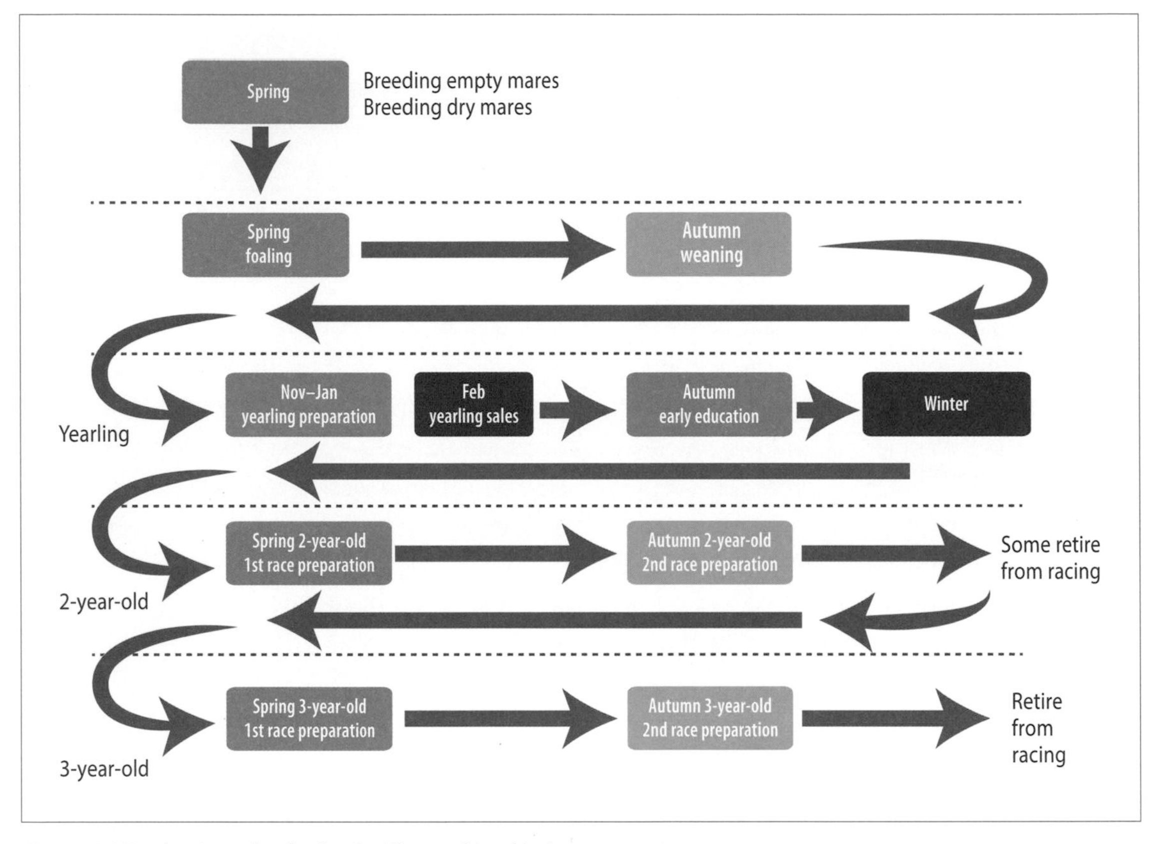

Figure 8.3 Production calendar for the Thoroughbred industry.

drugs, usually 18–21 days after foaling) or will be bred during the foal heat (the first heat 10–14 days after foaling); which method is used is dependent on the stud master and the condition of the mare after foaling.

In part because of breeding drift, most commercial breeders budget on mares producing three foals every four years. The majority of mares in the commercial breeding herd are less than 12 years old, as by this age they will have had at least two foals at racing age that have entered race training and will have demonstrated their merit as broodmares. To reflect this situation for tax purposes, in New Zealand a mare can be fully depreciated (written down to a zero book value) when she is 12 years old.

Most racing stallions are recruited to breeding after completion of their three- or four-year-old racing season. To be commercially successful, a stallion must have won a group one race; he will effectively have black type. The commercial life of a stallion in the racing industries is shorter than that of a mare, with the median commercial breeding career being 3–4 years (Rogers et al., 2016). This short commercial career is driven by the need for a stallion to have produced two-year-old winners from his first breeding season to ensure that breeders will continue to send mares to that stallion. The longevity of a stallion's breeding career depends on his ability to produce horses that will win major classic races (black type races).

Training Standardbreds on the beach.

The breeding of sport horses

The breeding of sport horses is not restricted, and there are a variety of breed organisations willing to register the resultant progeny. The lack of integration in sport horse breeding limits the recognition of successful sires and thus the genetic improvement of the national sport horse population (Creagh et al., 2010, 2012). The sport horse breeding industry could be considered semi-commercial, as some breeders generate a livelihood from breeding and producing sport horses (usually for the export market) while the majority breed horses for their own use or as a hobby.

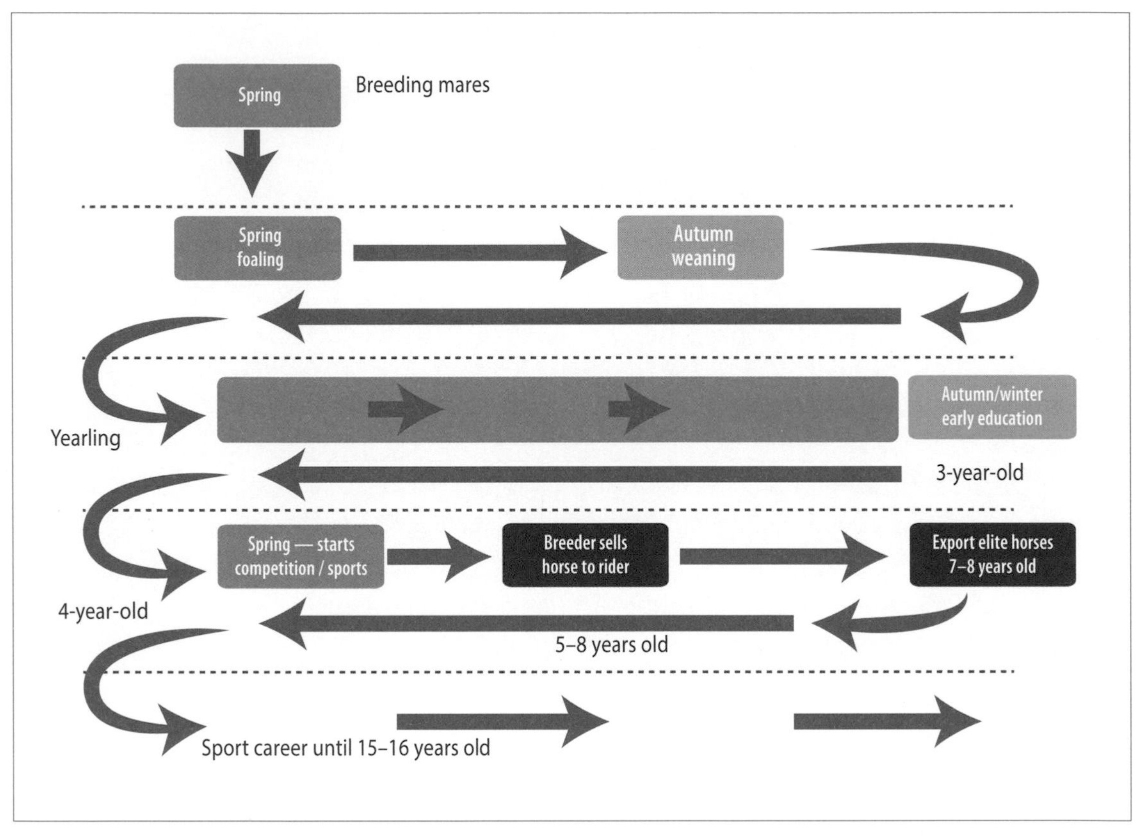

Figure 8.4 Production calendar for the sport horse industry.

The commercial stallions will have book sizes (the number of mares bred to a stallion in a season) similar to that observed in the racing industries. However, the majority of stallions will cover relatively few mares (fewer than seven mares per breeding season); most broodmare owners have 1–2 mares (George et al., 2013). This pattern of the majority of participants being hobby breeders is also observed in the sport horse industry within Europe.

The focus on marginal utility rather than economic return may explain the low rate of syndication of stallions (multiple owners with small shares in a stallion) in the New Zealand sport horse industry compared with the high frequency of its use in the racing industries.

Breeding calendar for sport horses

In part due to the structure of the industry (hobby rather than commercial breeding) and the later start of competitive life (usually four years for a sport horse), the breeding calendar (Figure 8.4) has less of a focus on the production of an early foal and maximisation of the reproductive output from the sire and the mare. In contrast to the racing industries, the breeding of sport horses occurs later in the season at a time that is close to the physiological norm for the horse. The smaller book size of the stallions and the lack of restriction on reproductive techniques also provides greater flexibility. The freedom to use AI has resulted in the extensive use of imported semen from Europe to breed show jumping and dressage horses.

Feral horses

A number of populations of feral horses exist in New Zealand, with clusters of herds in Northland, the East Coast and the central plateau of the North Island (Stafford, 2013). The feral horses of the central plateau are the most well-known population and are referred to as Kaimanawa horses (although they are generally only pony-sized, less than 14.2 hands high).

Due to the lack of natural predators to maintain a cap on the Kaimanawa horse population, the Department of Conservation conducts a biannual muster and selective cull of the feral herd to maintain numbers at around 300 horses, to prevent damage to the delicate sub-alpine ecosystem of the region.

At each biannual muster suitable horses, generally young stock, are offered for re-homing, for which there now appears to be a ready market as a number have been successful on the pony club competition circuit. Older stallions and mares not suitable for re-homing are sent to a local abattoir.

Wild horses on St James Station, North Canterbury.

Disease

The major disease, or disease risk, for horses — irrespective of industry sector — is musculoskeletal injury. The most robust data on musculoskeletal injury and lameness are from the Thoroughbred racing industry. Across the age groups, two-year-old horses have the greatest number of days unable to train, for which the primary reason is dorsal metacarpal disease (bucked shins or shin soreness) (Bolwell et al., 2012; Perkins et al., 2005a). This condition relates to localised remodelling and bone growth on the dorsal surface of the third metacarpal bone (the cannon bone) which causes an acute localised pain response to palpation, short striding and sometimes lameness. A major risk for developing dorsal metacarpal disease is a rapid increase in gallop work and insufficient exposure to high-speed work early in race training (Verheyen et al., 2005).

In older racehorses (four years and older), the injury profile shifts towards cyclic overload injuries, specifically tendon injury and strain or rupture of the superficial digital flexor tendon (SDFT) (Perkins et al., 2005b). The higher incidence of SDFT in older horses is partly due to changes in the structure of the tendon (a reduction in crimp angle) with age, leading to reduced elasticity of the tendon. Fatigue and sudden changes in ground conditions (soft going) are risk factors; combined with the age risk factor, they may contribute to the higher incidence of SDFT injuries seen in jumps-racehorses and event horses.

New Zealand is relatively free of many of the endemic equine diseases encountered in Europe and North America. The close economic and regulatory relationship of the equine industry with the Australian equine industry has seen close alignment of many of the regulations in trade and sport management. Both New Zealand and Australia are free of equine influenza and have an active national/international-level biosecurity programme to maintain this disease-free status. This status is unique internationally as equine influenza is endemic in all other countries except Iceland.

As there are few significant equine infectious diseases endemic in New Zealand, general property-level biosecurity is very lax, or non-existent, compared with similar European operations (Rosanowski et al., 2013a). At a national level, strangles (or equine distemper, caused by the bacterium *Streptococcus equi* var. *equi*) is the major endemic infectious disease; sporadic outbreaks occur each breeding season and at the start of the racing season. The clustering of these outbreaks at this time of the year is due to the increased frequency of horse movements (mares going to studs, and young horses entering racing stables or pre-training establishments) and the stressors of transport and new management systems (Rosanowski et al., 2013b, 2015).

Strangles is a difficult endemic disease to manage, as a number of horses may be carriers and/or shedding bacteria but not showing clinical signs. Attention to good biosecurity practices and judicious use of vaccines and antibiotics when indicated may aid in controlling the disease during outbreaks.

Racing and sport horses may be regularly vaccinated for tetanus and perhaps strangles and respiratory forms of equine herpesvirus (EHV). Broodmares are regularly vaccinated for tetanus, strangles and EHV-1 during pregnancy, and sometimes for salmonella. Hyper-immune plasma for the bacterium *Rhodococcus equi* (which causes pneumonia in foals) may be used on farms where the disease has been reported. The use of hyper-immune plasma is associated with larger stud farms and a greater density of mares and young stock (Rogers et al., 2007).

Animal welfare

The racing industries have the largest public profile, and as such are vulnerable to media scrutiny. Internationally, the strongest concerns about welfare and racing have centred on the issues of whip use, two-year-old racing and catastrophic musculoskeletal injury, particularly within jumps racing.

In accordance with international changes in the use of the whip in Thoroughbred racing, New Zealand adopted the use of the padded whip in 2009. In contrast to the debate in Australia and the United Kingdom around how the whip may be used during a race, there has been little such debate in New Zealand. The current rules covering whip use are relatively pragmatic and are strictly enforced by the Racing Integrity Unit, an independent organisation with responsibility for integrity issues in the New Zealand racing industry.

Within New Zealand, most racehorses (both Thoroughbred and Standardbred) are pre-trained and have initial race training as two-year-olds. While some will have a race start as a two-year-old, most will not do so until a three-year-old. Many trainers provide initial training for two-year-olds to develop the musculoskeletal system as well as for education (Bolwell et al., 2010b). There is now a significant body of data to demonstrate the positive effects of early training on development of the musculoskeletal system and on racehorse career length and success (Rogers et al., 2012).

Injury, and catastrophic musculoskeletal injury (CMSI) in particular, has the greatest profile in relation to the welfare of racehorses. Internationally, the rate of Thoroughbred CMSI is around two per 1000 starters for flat racing and six per 1000 starters for jumps racing. In New Zealand, the figure quoted for Thoroughbred flat racing CMSI is approximately 0.4 per 1000 starters (Tanner et al., 2016).

This low rate of CMSI is related to the low failure-to-finish rate and the consistency of the racing environment within New Zealand. The major risk factors for CMSI relate to field size and the presence of tracks graded as fast (hard or less compliant tracks). In New Zealand, medium-sized fields are usual, with only 3 per cent of races on tracks rated as fast and a pattern of racing that permits strategic spelling (rest periods) for racehorses.

The ability to manage horses at pasture provides a positive aspect to the management and welfare of horses in New Zealand, as this system permits management close to the natural ecological niche. An example of this is the perceived low incidence of colic in pasture-managed horses in New Zealand compared with its incidence in the northern hemisphere. Horses at pasture are also able to move freely and exhibit normal behaviour, and thus there is a lower prevalence of stereotypic behaviours. Possibly because of the lower cost of management, the less severe impact of the most recent global financial crisis in New Zealand and a pragmatic approach to equine euthanasia, there has not been the associated incidence of dumping of horses (abandonment on public areas) that has been observed within some European countries.

Horse meat is not widely consumed in New Zealand and therefore has only a limited domestic market. At present there is only one abattoir processing horses for human consumption and it is focused on the export market. Around 1500 horses are processed annually for human consumption and the number has remained at this level for the past couple of years.

It is estimated that the majority of horses euthanased in New Zealand are slaughtered for or processed into pet meat. Pet-food processors are not required to provide data on horses processed, but expert opinion puts the figure at 1250 with a further 2000 estimated to be euthanased as dog food for harrier packs. Combining these figures implies that about 4700 horses are euthanased for pet food or human consumption annually. An unknown number of horses are euthanased and buried on the owner's property.

Many horse owners are pragmatic about euthanasia that might be required due to compromised quality of life or loss of economic worth. However, the shift from a production-based management approach to that centred around marginal utility (emotional investment, as with a companion animal) is accompanied by an increase in the proportion of geriatric (very elderly) horses within the population and thus to welfare issues relating to the effective and fair management of aged horses.

Within the literature it is recognised that often the veterinarian and the owner of a geriatric horse differ significantly in their assessment of the welfare state of that horse, indicating under-reporting and poor recognition of illness and welfare issues in aged horses (Ireland et al., 2012). Although reasonably well described in Europe, there is currently little information about the welfare issues of geriatric horses in New Zealand.

The welfare of horses is regulated under the Animal Welfare Act 1999, with the minimum standards of care outlined in the Code of Welfare for Horses and Donkeys (NAWAC, 2006). This code provides pragmatic husbandry/management guides based on best practice and scientific research. The issuing of instant fines for minor incidences of horse abuse (e.g. inappropriate use of a whip) has recently been proposed. Prosecution of cases under the Animal Welfare Act 1999 are brought to court by either the Ministry for Primary Industries or the Society for the Prevention of Cruelty to Animals (SPCA).

Eventing.

References

Aurich, J., & Aurich, C. (2006). Developments in European horse breeding and consequences for veterinarians in equine reproduction. *Reproduction in Domestic Animals*, 41, 275–79.

Bolwell, C.F., Rogers, C.W., Firth, E.C., & French, N.P. (2010a). Management and exercise of Thoroughbred yearlings during preparation for yearling sales: a cross-sectional survey. *Proceedings of the New Zealand Society of Animal Production*, 70, 157–61.

Bolwell, C.F., Rogers, C.W., French, N.P., & Firth, E.C. (2012). Risk factors for interruptions occurring before the first trial start of 2-year-old Thoroughbred racehorses in training. *New Zealand Veterinary Journal*, 60, 241–46.

Bolwell, C.F., Rogers, C.W., Gee, E.K., & Rosanowski, S.M. (2014a). Descriptive statistics and the pattern of horse racing in New Zealand: Part 1 Thoroughbred racing. *Animal Production Science*. Accessed at http://dx.doi.org/10.1071/AN13442.

Bolwell, C.F., Rogers, C.W., Gee, E.K., & Rosanowski, S.M. (2014b). Descriptive statistics and the pattern of horse racing in New Zealand: Part 2 Standardbred racing. *Animal Production Science*. Accessed at http://dx.doi.org/10.1071/AN13443.

Bolwell, C.F., Russell, L.J., & Rogers, C.W. (2010b). A cross-sectional survey of training practices of 2-year-old racehorses in the North Island of New Zealand. *Comparative Exercise Physiology*, 7, 37–42.

Castle, K. (2012). Investigating the genetic and genomic basis of osteochondrosis in Thoroughbred horses from Australia and New Zealand (PhD thesis). Sydney: University of Sydney.

Cohen, N.D. (1994). Causes of and farm-management factors associated with disease and death in foals. *Journal of the American Veterinary Medical Association*, 204, 1644–51.

Costello, P., & Finnegan, P. (1988). *Tapestry of turf. The history of New Zealand racing 1840–1987*. Auckland: Moa Publications.

Creagh, F., Hickson, R.E., & Rogers, C.W. (2010). Preliminary examination of sport horse competition data for genetic evaluation. *Proceedings of the New Zealand Society of Animal Production*, 70, 143–45.

Creagh, F., Lopez-Villalobos, N., Hickson, R.E., & Rogers, C.W. (2012). Examination of New Zealand sport horse performance records and their suitability for the calculation of breeding values. *Proceedings of the New Zealand Society of Animal Production*, 72, 3–7.

Dicken, M., Gee, E.K., Rogers, C.W., & Mayhew, I.G. (2012). Gestation length and occurrence of daytime foaling of Standardbred mares on two stud farms in New Zealand. *New Zealand Veterinary Journal*, 60, 42–46.

Dijkstra, A., Sinnige, T.C., Rogers, C.W., Gee, E.K., & Bolwell, C.F. (2016). Preliminary examination of farriery and hoof care practices and owner-reported injuries in sport horses in New Zealand. *Journal of Equine Veterinary Science*, 46, 82–88.

Dykgraaf, S., Firth, E.C., Rogers, C.W., & Kawcak, C.E. (2008). Effects of exercise on chondrocyte viability and subchondral bone sclerosis in the distal third metacarpal and metatarsal bones of young horses. *Veterinary Journal*, 178, 53–61.

Fernandes, K.A., Bolwell, C., Gee, E., Rogers, C.W., & Thomas, D. (2014). A cross-sectional survey of rider and horse demographics, and the feeding, health and management of Pony Club horses in New Zealand. *Proceedings of the New Zealand Society of Animal Production*, 74, 11–16.

Fernandes, K.A., Rogers, C.W., Gee, E.K., Bolwell, C.F., & Thomas, D. (2015). Body condition and morphometric measures of adiposity in a cohort of Pony Club horses and ponies in New Zealand. *Proceedings of the New Zealand Society of Animal Production*, 75, 195–99.

Friedrich, C., Konig, S., Rogers, C.W., & Borstel, U.K.V. (2011). Examination of longevity in dressage horses — a comparison between Sport Horses in New Zealand and Hanoverians in Germany. *Zuchtungskunde*, 83, 68–77.

George, H.E.A., Bolwell, C.F., Rogers, C.W., & Gee, E.K.

(2013). A cross-sectional survey of New Zealand sport horse stud farms. *Proceedings of the New Zealand Society of Animal Production*, 73, 71–75.

Grace, N., Shaw, H.L., Gee, E.K., & Firth, E.C. (2002a). Determination of the digestible energy intake and apparent absorption of macroelements in pasture-fed lactating Thoroughbred mares. *New Zealand Veterinary Journal*, 50, 182–85.

Grace, N.D., Gee, E.K., Firth, E.C., & Shaw, H.L. (2002b). Digestible energy intake, dry matter digestibility and mineral status of grazing New Zealand Thoroughbred yearlings. *New Zealand Veterinary Journal*, 50, 63–69.

Grace, N.D., Rogers, C.W., Firth, E.C., Faram, T.L., & Shaw, H.L. (2003). Digestible energy intake, dry matter digestibility and effect of increased calcium intake on bone parameters of grazing Thoroughbred weanlings in New Zealand. *New Zealand Veterinary Journal*, 51, 165–73.

Hirst, R.L. (2011). Seasonal variation of pasture quality on commercial equine farms in New Zealand (Master's thesis). Palmerston North: Massey University.

Hoskin, S.O., & Gee, E.K. (2004). Feeding value of pastures for horses. *New Zealand Veterinary Journal*, 52, 332–41.

Ireland, J.L., Clegg, P.D., Mcgowan, C.M., Mckane, S.A., Chandler, K.J., & Pinchbeck, G.L. (2012). Comparison of owner-reported health problems with veterinary assessment of geriatric horses in the United Kingdom. *Equine Veterinary Journal*, 44, 94–100.

Liljenstolpe, C. (2009). *Horses in Europe*. Uppsala, Sweden: Swedish University of Agricultural Science.

Mincham, C. (2011). *The horse in New Zealand*. Auckland: David Bateman.

Morel, P.C.H., Bokor, A., Rogers, C.W., & Firth, E.C. (2007). Growth curves from birth to weaning for Thoroughbred foals raised on pasture. *New Zealand Veterinary Journal*, 55, 319–25.

Morley, P.S., & Townsend, H.G.G. (1997). A survey of reproductive performance in Thoroughbred mares and morbidity, mortality and athletic potential of their foals. *Equine Veterinary Journal*, 29, 290–97.

National Animal Welfare Advisory Committee (NAWAC). (2006). Code of Welfare: Horses and Donkeys. Wellington: Ministry for Primary Industries. Accessed at www.mpi.govt.nz/protection-and-response/animal-welfare/codes-of-welfare/.

National Research Council. (2007). *Nutrient requirements of horses*. Washington DC: National Academies Press.

Pearce, S.G., Firth, E.C., Grace, N.D., & Fennessy, P.F. (1998a). Effect of copper supplementation on the evidence of developmental orthopaedic disease in pasture-fed New Zealand Thoroughbreds. *Equine Veterinary Journal*, 30, 211–18.

Pearce, S.G., Grace, N.D., Firth, E.C., Wichtel, J.J., Holle, S.A., & Fennessy, P.F. (1998b). Effect of copper supplementation on the copper status of pasture-fed young Thoroughbreds. *Equine Veterinary Journal*, 30, 204–10.

Perkins, N.R., Reid, S.W.J., & Morris, R.S. (2005a). Profiling the New Zealand Thoroughbred racing industry. 2. Conditions interfering with training and racing. *New Zealand Veterinary Journal*, 53, 69–76.

Perkins, N.R., Reid, S.W.J., & Morris, R.S. (2005b). Risk factors for injury to the superficial digital flexor tendon and suspensory apparatus in Thoroughbred racehorses in New Zealand. *New Zealand Veterinary Journal*, 53, 184–92.

Randall, L., Rogers, C.W., Hoskin, S.O., Morel, P.C.H., & Swainson, N.M. (2014). Preference for different pasture grasses by horses in New Zealand. *Proceedings of the New Zealand Society of Animal Production*, 74, 5–10.

Rogers, C., & Wickham, G. (1993). Studies of alternative selection policies for the New Zealand sport horse. *Proceedings of the New Zealand Society of Animal Production*, 53, 423–26.

Rogers, C.W., Bolwell, C.F., & Gee, E.K. (2012). Proactive management of the equine athlete. *Animals*, 2, 640–55.

Rogers, C.W., Bolwell, C.F., & Gee, E.K. (2014a). Early exercise in the juvenile horse to optimise performance later in life. In A. Paz-Silva, M.S.A. Vazquez & R.S.-A. Fernandez (Eds), *Horses: breeding, health disorders, and effects on performance and behavior*. New York: Nova Publishing.

Rogers, C.W., Bolwell, C.F., Gee, E.K., Peterson, M.L., & Mcilwraith, C.W. (2014b). Profile and surface conditions of New Zealand Thoroughbred racetracks. *Journal of Equine Veterinary Science*, 34, 1105–09.

Rogers, C.W., & Firth, E.C. (2005). Preliminary examination of the New Zealand event horse production system. *Proceedings of the New Zealand Society of Animal Production*, 65, 372–77.

Rogers, C.W., Firth, E.C., Mcilwraith, C.W., Barneveld, A., Goodship, A.E., Kawcak, C.E., Smith, R.K.W., & Van Weerew, P.R. (2008). Evaluation of a new strategy to modulate skeletal development in Thoroughbred performance horses by imposing track-based exercise during growth. *Equine Veterinary Journal*, 40, 111–18.

Rogers, C.W., & Gee, E.K. (2011). Selection decisions in Thoroughbred broodmares. *Proceedings of the New Zealand Society of Animal Production*, 71, 122–25.

Rogers, C.W., Gee, E.K., & Bolwell, C.F. (2016). Reproductive production constraints within the New Zealand racing industry. *Proceedings of the New Zealand Society of Animal Production*, 76, 146–50.

Rogers, C.W., Gee, E.K., & Firth, E.C. (2007). A cross-sectional survey of Thoroughbred stud farm management in the North Island of New Zealand. *New Zealand Veterinary Journal*, 55, 302–07.

Rogers, C.W., Gee, E.K., & Vermeij, E. (2009.) Retrospective examination of the breeding efficiency of the New Zealand Thoroughbred and Standardbred. *Proceedings of the New Zealand Society of Animal Production*, 69, 126–31.

Rogers, C.W., & Vallance, A. (2009). Studbook application for membership of the World Breeding Federation for Sport Horses. New Zealand Sport Horse Promotion Board Inc.

Rosanowski, S.M. (2012). Epidemiological investigations of the New Zealand horse population and the control of equine influenza (PhD thesis). Palmerston North: Massey University.

Rosanowski, S.M., Cogger, N., Rogers, C.W., Benschop, J., & Stevenson, M.A. (2013a). An investigation of the movement patterns and biosecurity practices on Thoroughbred and Standardbred stud farms in New Zealand. *Preventive Veterinary Medicine*, 108, 178–87.

Rosanowski, S.M., Rogers, C.W., & Cogger, N. (2015). The movement pattern of horses around race meetings in New Zealand. *Animal Production Science*, 55, 1075–80.

Rosanowski, S.M., Rogers, C.W., Cogger, N., Benschop, J., & Stevenson, M.A. (2012). A description of the demographic characteristics of the New Zealand non-commercial horse population with data collected using a generalised random-tessellation stratified sampling design. *Preventive Veterinary Medicine*, 107, 242–52.

Rosanowski, S.M., Rogers, C.W., Cogger, N., Stevenson, M.A., & Benschop, J. (2013b). Analysis of horse movements from non-commercial horse properties in New Zealand. *New Zealand Veterinary Journal*, 61, 245–53.

Simmons, H.R.G. (2015). Investigation of current equine identification systems utilised in the New Zealand equine industry. Palmerston North: Massey University.

Stafford, K. (2013). Welfare of horses. In *Animal welfare in New Zealand*. Cambridge, New Zealand: New Zealand Society of Animal Production.

Stowers, N.L., Erdtsieck, B., Rogers, C.W., Taylor, T.B., & Firth, E.C. (2010). The prevalence of limb deformities in New Zealand Standardbred foals and their influence on racing sucess — a preliminary investigation. *Proceedings of the New Zealand Society of Animal Production*, 70, 140–42.

Stowers N.L., Rogers, C.W., & Hoskin, S.O. (2009). Management of weanlings on commercial Thoroughbred stud farms in the North Island of New Zealand. *Proceedings of the New Zealand Society of Animal Production*, 69, 4–9.

Tanner, J.C., Rogers, C.W., Bolwell, C.F., Cogger, N., Gee, E.K., & Mcilwraith, C.W. (2016). Analysis of failure to finish a race in a cohort of Thoroughbred racehorses in New Zealand. *Animals*, 6, doi:10.3390/ani6060036.

Tanner, J.C., Rogers, C.W., Bolwell, C.F., & Gee, E.K. (2012). Preliminary examination of wastage in Thoroughbred and Standardbred horses in New Zealand using training milestones. *Proceedings of the New Zealand Society of Animal Production*, 72, 172–74.

Tanner, J.C., Rogers, C.W., & Firth, E.C. (2011). The relationship of training milestones with racing success in a population of Standardbred horses in New Zealand. *New Zealand Veterinary Journal*, 59, 323–27.

Tanner, J.C., Rogers, C.W., & Firth, E.C. (2013). The association of 2-year-old training milestones with career length and racing success in a sample of Thoroughbred horses in New Zealand. *Equine Veterinary Journal*, 45, 20–24.

Van Der Heyden, L., Lejeune J.P., Caudron, I., Detilleux, J., Sandersen, C., Chavatte, P., Paris, J., Deliege, B., & Serteyn, D. (2013). Association of breeding conditions with prevalence of osteochondrosis in foals. *Veterinary Record*, 172, 68–71.

Verhaar, N., Rogers, C.W., Gee, E., Bolwell, C., & Rosanowski, S.M. (2014). The feeding practices and estimated workload in a cohort of New Zealand competition horses. *Journal of Equine Veterinary Science*, 34, 79–84.

Verheyen, K.L.P., Henley, W.E., Price, J.S., & Wood, J.L.N. (2005). Training-related factors associated with dorsometacarpal disease in young Thoroughbred racehorses in the UK. *Equine Veterinary Journal*, 37, 442–48.

Waldron, K., Rogers, C.W., Gee, E.K., & Bolwell, C.F. (2011). Production variables influencing the auction sales price of New Zealand Thoroughbred yearlings. *Proceedings of the New Zealand Society of Animal Production*, 71, 92–95.

Williamson, A., Rogers, C.W., & Firth, E.C. (2007). A survey of feeding, management and faecal pH of Thoroughbred racehorses in the North Island of New Zealand. *New Zealand Veterinary Journal*, 55, 337–41.

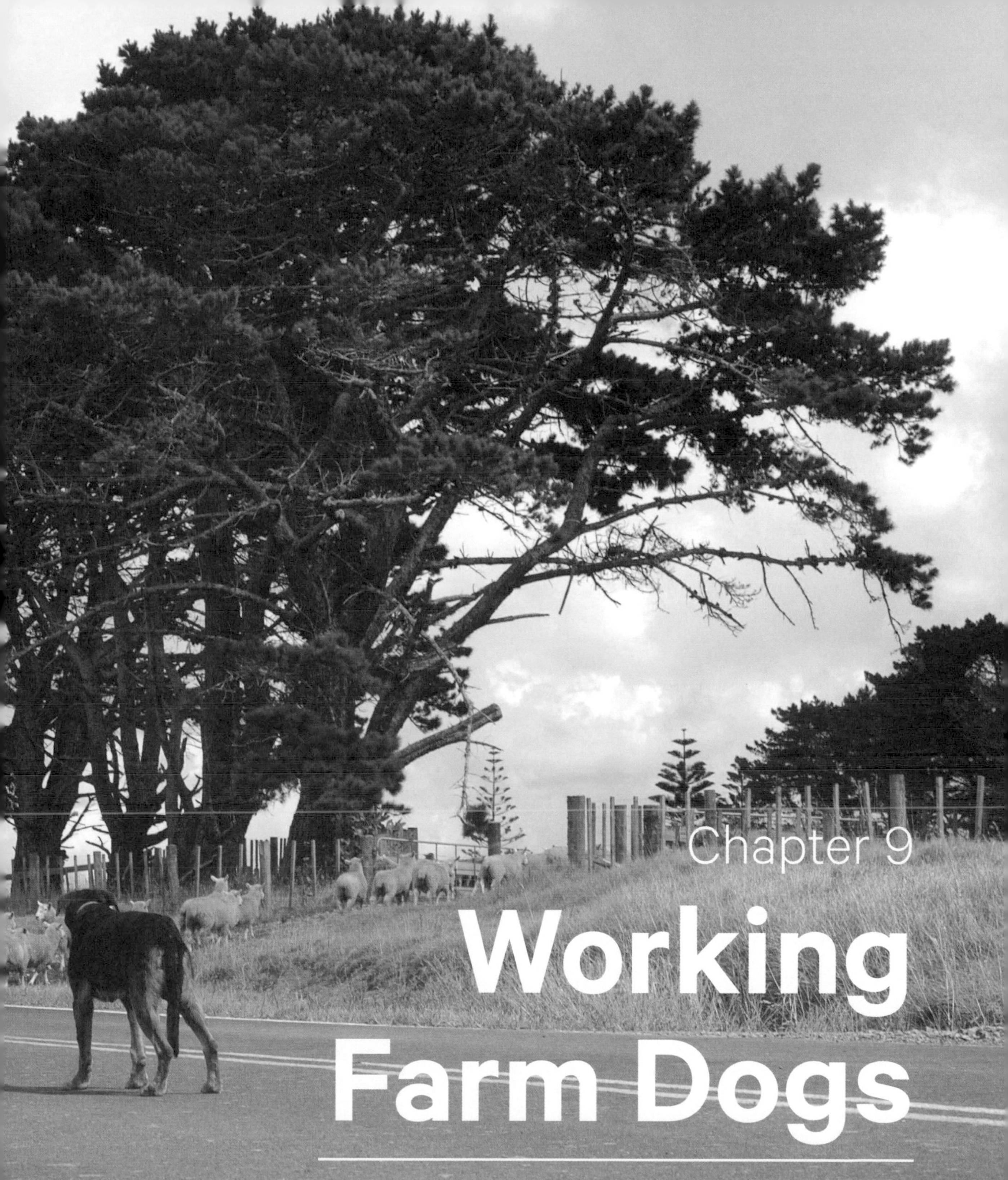

Chapter 9

Working Farm Dogs

Naomi Cogger and Helen Sheard

Chapter 9

Working Farm Dogs

Naomi Cogger and Helen Sheard
Institute of Veterinary, Animal and Biomedical Sciences
Massey University, Palmerston North

Introduction

The first major influx of sheep into New Zealand occurred in 1834, when Merinos from Australia were landed on Mana Island before being moved to the Wairarapa. From 1840 on, more sheep were imported from Australia to graze arable land that was rapidly expanding as bush and scrub were cleared (Dalton & Orr, 2004). It is thought that the original working farm dogs were brought to New Zealand by shepherds in the mid-1800s, when sheep numbers were increasing (Hughes, 2013). The first sheep dog trials in New Zealand were recorded at Wanaka in 1867, at Waitangi in 1868, and at Haldon Station in the Mackenzie Country in 1870.

There is limited documentation on the development of working farm dogs in New Zealand. Special-interest books suggest that the first working farm dog breeds in New Zealand were Border Collies that arrived with Scottish settlers, and Bearded Collies that were brought out by English settlers (Dalton, 1996; Oliver et al., 2004; Redwood, 1980; Rennie, 1984). Other breeds that are thought to have arrived in New Zealand during early European settlement, albeit in smaller numbers, are the Smithfield and Scottish 'hunter' dog (Rennie, 1984).

Today, New Zealand has over 55,000 farming operations, of which approximately 25,000 are classified as predominantly sheep and beef (Beef & Lamb New Zealand, 2015), and an estimated 200,000 working farm dogs. There is no doubt that these dogs make an economic contribution to sheep and beef farming, as mustering extensive areas of hill and high country would be difficult, if not impossible, without them. To put their value in context, a study of Australian working dogs has estimated that if a dog were to be paid a salary for the hours worked, it would have a median lifetime earning of AUD40,000 (NZ$44,000), which represents a 5.2-fold return on investment (Arnott et al., 2014). It would be reasonable to assume that dogs in New Zealand would earn at least the same amount; in fact, the economic value would exceed $44,000 if one was to take into consideration the labour units that are replaced by a single working dog.

Types of working farm dog

Farmers in New Zealand categorise dogs in terms of natural instinct rather than breed. The three main types of working dog on New Zealand sheep and beef farms are the huntaway, heading dog and handy dog — none of which are registered breeds with the New Zealand Kennel Club. Occasionally, dogs not traditionally considered to be working dogs are used to work stock; examples are Labrador Retrievers, Jack Russells and Fox Terriers (Jerram, 2013).

The huntaway is unique to New Zealand; it has been selectively bred to enhance the dog's natural instinct to 'hunt' or chase sheep away by barking. Huntaways are large dogs that weigh between 30 and 40 kg, are usually black-and-tan coloured and have coats that are either smooth or long and shaggy; there is, however, considerable variation in their physical appearance.

The exact origins of the huntaway are uncertain; there is some speculation that the original breeds playing a role in producing the classic black-and-tan barking dog included the Border Collie with Gordon Setter added for barking ability. Bearded Collies, German Shepherds, black Labradors, Smithfield (bobtail) Collies and Foxhounds may also contribute to the

Examples of variation in the physical appearance of the New Zealand huntaway. **Previous:** A huntaway keeps a careful eye on the mob.

genetic make-up (Hughes, 2013).

Heading dogs work silently and have a natural instinct to circle widely around stock. The dogs will work quietly close to the stock during lambing to separate an animal from the group. The majority of heading dogs are described as New Zealand heading dogs, which is not a registered breed. New Zealand heading dogs come from the British Border Collie and, like the huntaway, there is considerable variation in colour, size and coat. Typically, the New Zealand heading dog weighs around 20 kg, is black and white or black, tan and white in colour, and has a smooth coat.

Handy dogs are capable of doing both heading and huntaway work. The dogs may be a cross between the huntaway and heading dogs or other working dog breeds. This type of dog has no specific appearance.

Huntaways and heading dogs are the most common farm working dogs (Cave et al., 2009; Jerram, 2013; Singh et al., 2011). In a survey of working dogs presenting at veterinary clinics, huntaways accounted for 50 per cent of the dogs and heading dogs accounted for 40 per cent (Cave et al., 2009). A similar ratio of huntaways and heading dogs was reported in a survey of farms in the lower half of the North Island (Jerram, 2013). In contrast, a survey of the members of the New Zealand Sheep Dog Trial Association reported that approximately 50 per cent were heading dogs and 40 per cent were huntaways. A possible reason for the difference is that those involved in trialling sheep dogs may have a preference for heading dogs; alternatively, those who trial dogs may be more likely to work stock on farms for which heading dogs are more useful than huntaways.

Dog work

About 85 per cent of dogs are worked solely by their owner, with the remainder worked by their owner and other people on the farm (Cogger et al., 2016). Typically, a farmer (or shepherd) will have a team of between four and six dogs that are a mixture of huntaway, heading and handy. The number of dogs on a farm will increase with farm size and steepness of contour. However, the number of dogs that a single owner has does not vary. On larger hill-country or high-country farms, there are more workers with dogs rather than the owners having more dogs.

Regardless of type, dogs are used to work both sheep and cattle. Trials are only performed with sheep, and most of the descriptions relate to using dogs for sheep work. All types of dog can be used to move, or muster, stock, although they will do so in different ways. As noted previously, a huntaway will bark and work behind to 'hunt' the sheep away. In contrast, heading dogs will work at the front, leading the sheep and keeping them in check if need be.

Heading dogs can also be used to cut an animal from a mob, which can be useful at lambing time. Some heading dogs will show an instinct to hold sheep, which allows them to contain a mob of sheep quietly while they wait for the farmer to arrive. A huntaway or handy is particularly useful as a yard dog and in loading or unloading sheep. In the yards, a dog will run over the backs of sheep

The team at work shifting ewes and lambs — dogs under control.

New Zealand heading dog working in stock yards.

and walk through a mob to keep stock moving.

To better understand the working life of dogs, researchers conducted a survey of dog owners on sheep and beef farms in the lower half of the North Island (Cogger et al., 2016; Jerram, 2013). Information was gathered about dogs more than six months of age that were on the farm at the time of the visit or had been on the farm in the previous 12 months. In total, 1117 dogs were included in the study of which 754 dogs were considered fully trained, 290 partially trained and 71 retired; two had no career information recorded.

Of those that were fully trained, 87 per cent were present on the day the researcher visited the farm, 10 per cent were dead and the remainder had been sold or given away. Of the 168 dogs that were dead at the time of the farm visit, 22 per cent had died in the natural course of life, 58 per cent had been euthanased by the owner and 20 per cent had been euthanased by a veterinarian.

It is noteworthy that the classification of the dogs into the three career stages proved difficult. First, there was difficulty in reaching a definition of retirement that could be applied consistently across farms. It was often reported that dogs in semi-retirement would still go out each day with the team but would not do as much work. Second, with dogs receiving 'on the job' training it was difficult to determine exactly when a dog had become fully trained. Consequently, future research will need to develop definitions and criteria to classify workload that take into consideration the fact that (unlike other working dogs) those on farms 'ease' into both work and retirement.

Training

Approximately 10 per cent of working farm dogs are acquired fully trained and a further 5 per cent partially trained (Jerram, 2013). The remaining 85 per cent are bred by the owner or purchased untrained. The average age for the commencement of training is six months. The majority of owners conduct the training while working stock, rather than running specific training sessions.

Housing

Overnight, farm dogs are housed in purpose-built shelters rather than in the owners' residence (Jerram, 2013). Over 80 per cent of these shelters have been built by the owner rather than commercially produced. Owner-built housing is variable. In contrast, commercially produced housing tends to have elevated boxes with a caged run attached and wooden flooring. Surveys conducted in the North Island found that bedding is supplied in approximately 20 per cent of cases and may take the form of woollen blankets, carpet, a sack, a duvet, clothing, straw, sheep wool, a commercially produced dog hammock, sections of a foam mattress, or a vehicle foot-mat.

Of particular importance in the housing of working farm dogs is ensuring that the ambient temperature does not drop below a level that will require the dog to increase its metabolic rate in order to produce heat (the lower critical temperature). Studies in non-working dogs indicate that the lower critical temperature for dogs is between 20°C and 25°C (Subcommittee on Dog and Cat Nutrition, 2006).

While no studies have investigated the ambient temperature in different housing conditions, overnight temperatures in New Zealand winters are well below 20°C. It would therefore be reasonable to conclude that ambient temperatures in many scenarios are sufficiently low that the dogs would need to increase their metabolism in order to regulate body temperature. Consequently, shelters should be built with adequate insulation and/or heating to ensure that ambient temperatures do not drop below 20°C.

Commercially produced housing for working farm dogs.

Huntaway rugged up against the cold to prevent thermal loss overnight.

Nutrition

Energy

The energy requirements of working farm dogs have not been determined. However, researchers have suggested that the energy requirements are between 1.5 and 3 times those of pet dogs (Cave, 2009; Guilford, 1997; Singh et al., 2011). The exact increase will depend on a range of factors such as the type of work, the topography of the farm and the ambient temperature.

The high degree of variation both between dogs and throughout the year means that the feeding of individual dogs should focus on maintaining body condition (Figure 9.1). That said, the optimal body condition score for working dogs has not been determined either, although this is an area of active research.

Macronutrients

The ratio of fat, protein and carbohydrate available can alter the energy source selected during exercise and as such may affect a dog's performance. At rest, muscles preferentially use fat (lipids) as their fuel source. The most efficient of these lipids are non-esterified fatty acids that are metabolised to release energy relatively slowly. When the dog's exercise intensity increases, glucose becomes the predominant source of energy (Cave, 2009). For example, greyhounds are reliant on carbohydrate, which is stored in the muscle as glycogen and can be quickly metabolised for sprinting. Dogs used for endurance purposes preferentially use free fatty acids for muscular work. High-fat (> 46 per cent), low-carbohydrate (26–28 per cent) diets increase the storage of fat within muscle, and also the rate of fat usage, thereby increasing endurance through preserving muscle glycogen stores. Feeding high-carbohydrate diets can also increase glycogen storage in muscle; however, a high carbohydrate intake also increases the rate at which carbohydrate is metabolised, thus negating its effect. Muscle glycogen is thus preserved more effectively by feeding high-fat diets (Cave, 2009), and it would appear that the ideal working dog diet would have high fat and low carbohydrate levels.

Working dogs will also have a higher rate of muscle catabolism (breakdown). Therefore, working dogs are likely to have a higher requirement for dietary protein than sedentary non-working dogs. Higher protein in the diet may also play a role in preventing both bone and soft-tissue injuries. A study by Reynolds et al. (1999) found that sled dogs with 19 per cent of metabolisable energy (ME) obtained from protein had eight times more musculoskeletal injuries than those with 24 per cent of ME from protein. The timing of feeding may also affect the incidence of musculoskeletal injuries. Feeding a readily digestible protein source within two hours of exercise has been shown to promote recovery from muscle fatigue; and digestible carbohydrate sources given during or immediately after exercise improve endurance and promote greater muscle glycogen repletion (Cave, 2009).

Feeding practices

Farm dogs in New Zealand are typically fed once a day with a combination of commercial biscuits and home-kill (Jerram, 2013; Singh et al., 2011). The standard practice of feeding dogs once a day may make it physically impossible for the dog to consume enough food to meet its nutrient requirements, particularly in periods of high workload. Therefore, during such periods it would be advisable to alter the composition of the diet to include more energy-dense food. However, this does not appear to be standard practice with owners, who have reported that the amount of each component fed increased with increasing workload but the composition remained largely unchanged (Singh et al., 2011).

Offal from home-kill can spread the parasite *Echinococcus granulosus* that causes hydatids (tapeworm). While New Zealand has been declared provisionally free of this parasite, control notices are still in place and it is a legal requirement that all home-killed offal must be treated prior to feeding (Ministry for Primary Industries, 2010). The parasite *Taenia ovis*, also known as sheep measles, is still a concern and for this reason the non-profit organisation Ovis Management recommends that all sheep meat be treated prior to feeding. In both cases, treatment can be done by cooking or freezing. If cooking the meat or offal, it must be boiled for a minimum of 30 minutes. When using a freezing treatment, meat must be frozen at –10°C or less for a minimum of 10 days.

Feeding dogs a high level of farm-killed meat could result in a diet that is deficient in vitamins and minerals. Farm-kill mutton is likely to be deficient in or have marginal concentrations of a number of vitamins and minerals, such as iodine, copper, and vitamins A, E and B12 (Cave, 2009). Good-quality commercial dry diets are likely to compensate for the vitamin and mineral inadequacies of meat, as long as meat comprises no more than 30 per cent of the diet (Guilford, 1997).

If home-kill does not include bones, then diets may be deficient in calcium and phosphorus; however, feeding bones increases the risk of perforation or obstruction of the bowel, and constipation.

WSAVA Global Nutrition Committee

UNDER IDEAL

1. Ribs, lumbar vertebrae, pelvic bones and all bony prominences evident from a distance. No discernible body fat. Obvious loss of muscle mass.
2. Ribs, lumbar vertebrae and pelvic bones easily visible. No palpable fat. Some evidence of other bony prominences. Minimal loss of muscle mass.
3. Ribs easily palpated and may be visible with no palpable fat. Tops of lumbar vertebrae visible. Pelvic bones becoming prominent. Obvious waist and abdominal tuck.

IDEAL

4. Ribs easily palpable, with minimal fat covering. Waist easily noted, viewed from above. Abdominal tuck evident.
5. Ribs palpable without excess fat covering. Waist observed behind ribs when viewed from above. Abdomen tucked up when viewed from side.

OVER IDEAL

6. Ribs palpable with slight excess fat covering. Waist is discernible viewed from above but is not prominent. Abdominal tuck apparent.
7. Ribs palpable with difficulty; heavy fat cover. Noticeable fat deposits over lumbar area and base of tail. Waist absent or barely visible. Abdominal tuck may be present.
8. Ribs not palpable under very heavy fat cover, or palpable only with significant pressure. Heavy fat deposits over lumbar area and base of tail. Waist absent. No abdominal tuck. Obvious abdominal distention may be present.
9. Massive fat deposits over thorax, spine and base of tail. Waist and abdominal tuck absent. Fat deposits on neck and limbs. Obvious abdominal distention.

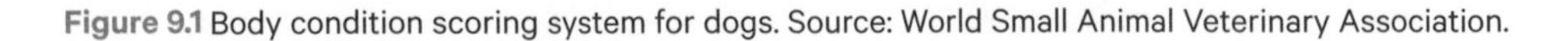

Figure 9.1 Body condition scoring system for dogs. Source: World Small Animal Veterinary Association.

Physical examination of working dog enrolled in Vetlife's TeamMate study.

Animal welfare

Working dogs are bred to be capable of moving stock. They are allowed the opportunity to engage in this behaviour in the context of a controlled work environment. The welfare issues affecting working dogs may relate to physical problems such as malnutrition, overwork, injury and ill-health, and poor housing conditions during winter. Injuries are common and may not be treated appropriately.

Other problems for some dogs include brutal treatment during training and work. Training using excessive punishment results in poor outcomes and probably more dog wastage than should be the case. Another source of wastage is that many more puppies are born than are needed as replacements; these may be killed by drowning, which is unacceptable. Fortunately, more research is currently being carried out on the dietary and physical requirements of working dogs and training techniques. The high value of a good working dog makes sensible and sensitive management of them good financial sense, and makes farm work a more pleasant activity for all.

Animal health

The fact that farm dogs work with stock, around vehicles, and have high activity levels places them at risk of injury and disease. Furthermore, working farm dogs are at increased risk of reproductive diseases and unnecessary breeding as most are sexually entire. Until recently, research on health in working dogs focused on specific diseases such as hip dysplasia (Hughes, 2001), and the prevalence of case reports resulted in a bias towards the unusual rather than the common.

More recently, studies have been undertaken to describe the range of health problems in working farm dogs presenting at veterinary hospitals (Cave et al., 2009) and to survey owner-reported health conditions (Jerram, 2013; Sheard, 2014). While these studies provide valuable information, their design produces biased estimates given that not all dogs with health problems are taken to veterinary clinics and owners will not detect all health problems.

To address these gaps, Vetlife, in collaboration with Massey University's Working Dog Centre, launched TeamMate in 2013. TeamMate is an observational study that will follow working farms dogs for five years; to date, over 500 dogs have been enrolled in the study. Working dogs are enrolled when they reach 18 months of age, at which time they are assessed by a veterinarian. The dogs are then assessed twice a year until the end of the study, death, retirement or until a dog is lost to follow-up. The assessment includes body weight, body condition score and morphometric measurements with a tape measure; trot up to check for lameness; visual inspection of coat and skin, ears and teeth, pads and nails to detect any signs of ill-health; cardiovascular and respiratory examination; and manual assessment (palpation) of legs, tail, muscles, joints, mammary glands, testes and lymph nodes. TeamMate will provide better information regarding the longevity and health of working dogs.

Trauma

Traumatic injuries are a common occurrence for working dogs, accounting for 38 per cent of all presentations to veterinary clinics in a 12-month period (Cave et al., 2009). In the survey of owner-reported health problems previously mentioned, 25 per cent of dogs had one or more traumatic injuries in the preceding 12 months (Sheard, 2014). The most common anatomical locations for traumatic injury seen at veterinary clinics were the foot (40 per cent), stifle (15 per cent; equivalent of the human knee) and tarsus (10 per cent; heel bone), and the most common causes of trauma were stock, automobiles and fences. In owner-reported events, the most commonly affected body systems were the musculoskeletal system (12 per cent) and the skin (12 per cent). While the nature of farm dog work is such that it will be difficult, if not impossible, to prevent all traumatic injuries, there is certainly room for improvement and this should be an area of active research.

Diseases of the gastrointestinal tract

Problems with the gastrointestinal (GI) tract were the most common non-trauma-related reason for presentation at a veterinary clinic, accounting for 9 per cent of visits (Cave et al.,

2009). In contrast, the survey of owner-reported conditions found that GI tract problems were the fifth most common complaint, accounting for 5 per cent of non-trauma-related health problems (Sheard, 2014). In both studies, constipation was the most common problem, followed by GDV (gastric dilatation–volvulus or twisted bowel) as reported by owners. Further analysis of the data from veterinary clinics found that huntaways were 17 times more likely than heading dogs to experience GDV and that the risk of GDV was higher in old dogs and during the summer months (Hendriks et al., 2012). Interestingly, when information was available about stomach contents most dogs had food, bones or other indigestible contents in the stomach; this finding differs from previous studies, where gas was the predominant stomach content (Hill, 2010).

Non-traumatic musculoskeletal disease

Degenerative joint disease was the most commonly reported musculoskeletal disease of working farm dogs, accounting for 12 per cent of non-trauma-related presentations at veterinary clinics (Cave et al., 2009) and reported as affecting 10 per cent of dogs in the survey of owner-reported health problems (Sheard, 2014). Both of these studies are likely to have underestimated the prevalence of degenerative joint disease — preliminary, and as yet unpublished, results from the TeamMate project show that approximately 25 per cent of working farm dogs have at least one joint with a reduced range of motion. The high prevalence of dogs with degenerative joint disease is of concern, as the condition can cause pain and may affect performance.

The other non-traumatic musculoskeletal condition of concern, particularly in huntaways, is hip dysplasia. A study involving 93 huntaways and 48 heading dogs presenting to a veterinary clinic for routine examination for problems other than lameness found that 24 per cent of the huntaways and 6 per cent of the heading dogs had hip dysplasia (Hughes, 2001). The average score for huntaways was 10.8, which made it the fifth worst of the breeds assessed by the New Zealand Veterinary Association scheme. It is noteworthy that the dogs had presented for conditions other than lameness and that none of the owners of dysplastic dogs had noticed any signs of lameness in their animals; this finding caused Hughes (2001) to postulate that radiological and physical signs may not correlate well in working dogs because they are typically lean, fit and highly motivated to work.

Reproductive disease

Conditions involving the reproductive tract account for 19 per cent of veterinary visits unrelated to trauma (Cave et al., 2009). In contrast, when relying on owner-reported events then conditions involving the reproductive tract were reported in only 3 per cent of dogs (Sheard, 2014). The most common reproductive conditions requiring veterinary treatment were pyometra/endometritis, vaginal prolapse and vaginal hyperplasia. Forty-two dogs were also treated for mis-mating.

Owner-reported conditions showed a slightly different pattern, with infertility the most common problem followed by mammary cancer; vaginal prolapse was relatively rare, occurring in less than 1 per cent of dogs. With most farm dogs being sexually entire, the prevalence of these conditions and of unnecessary mating could be reduced if the dogs were desexed as soon as the breeding of replacements had been accomplished or the animals had been determined to be of no value for breeding.

Infectious diseases

The close association between working dogs and sheep and cattle, poor and closely confined housing, and opportunities to eat raw meat puts working farm dogs at increased risk of infectious diseases. Those that have received the most attention in New Zealand working dogs are leptospirosis and intestinal parasites.

Leptospirosis

The six leptospiral serovars known to be present in New Zealand are *Leptospira interrogans* serovars Pomona and Copenhageni, and *L. borpetersenii* serovars Balcanica, Hardjo, Ballum and Tarassovi. Studies have shown that rural dogs are significantly more likely to be seropositive to Pomona and Copenhageni than urban dogs (O'Keefe et al., 2002). A more recent survey also found that Copenhageni was the most common leptospiral serovar isolated from farm working dogs, with a prevalence of 10 per cent (Cave et al., 2014). These researchers also found that rural dogs were more likely to be seropositive for Hardjo than urban dogs. While no studies have examined the prevalence of leptospirosis in the South Island, it seems unlikely that dogs are not infected given that the pathogen is present in sheep and cattle in the South Island.

It would be prudent for dogs on farms with leptospirosis present to receive appropriate vaccinations. That said, there is only one licensed vaccination for dogs and this only protects against Copenhageni (Cave et al., 2013). Recently, researchers have examined the serological response of dogs to a commercial vaccine for use in cattle containing cultured strains of *L. interrogans* serovars Copenhageni and Pomona, and *L. borgpetersenii* serovar Hardjo (Cave et al., 2014). The study showed that the vaccination can raise the level of antibodies to all three serovars but it is not known whether this increase is sufficient to guarantee protection. More work is therefore needed before the off-label use of the cattle vaccination can be recommended.

Intestinal parasites

In 2002, the MPI declared New Zealand as being provisionally free of the dog tapeworm *Echinococcus granulosus* that causes hydatids (MPI, 2010). However, *Taenia ovis*, also known as sheep measles, is still of concern and all dogs should be regularly treated for tapeworm using a regimen that has been developed in consultation with a veterinarian or animal health advisor (Ovis Management, n.d.).

A number of other nematode and/or protozoan parasites are also common in working farm dogs. In 2010, faecal samples taken from healthy working farm dogs found that 40 per cent had a nematode and/or protozoan parasite infection, of which more than 40 per cent were infected with more than one species (O'Connell et al., 2013). The most common nematode infections were hookworms (12 per cent), *Toxocara canis* (5 per cent) and *Trichuris vulpis* (5 per cent); the most common protozoan infections were by species of *Sarcocystis* (21 per cent), *Giardia* (8 per cent), *Isopora* (5 per cent) and *Neospora* (2 per cent).

It is not clear whether the high prevalence of parasites was unique to the region of New Zealand in which the study was conducted. Neither is it known why the prevalence of nematode and/or protozoan infection is so high in farm working dogs. A possible explanation is that dog owners are not administering anthelmintic drugs sufficiently regularly; a second possible cause is that the commonly used anthelmintic drugs are not efficacious.

Toxicity

Farm environments contain a number of chemicals, including herbicides, pesticides, stock feed additives and animal remedies (Parton, 2002). However, the most common poisoning observed by Cave et al. (2009) was due to anticoagulants — i.e. rat bait, a chemical that can be found in many households across New Zealand.

While not a common occurrence, working dog owners also need to be aware of macrocyclic lactone toxicity. This condition can occur via direct ingestion or skin absorption of the chemicals abamectin, ivermectin or moxidectin and the ingestion of faeces from livestock and horses treated with these chemicals. The chemicals are found in a number of sheep, beef and horse drenches; horse drenches in particular may have a greater risk of ingestion by dogs as many have a flavour additive that may be attractive.

Dogs may also be exposed when owners treat them with macrocyclic-lactone-containing drenches in the mistaken belief that these are safe (Parton et al., 2012). While macrocyclic lactone toxicity can occur in any dog, it is of particular concern for those carrying the ABCB1 mutant allele. Dogs with this mutation are sensitive to these compounds, with a toxic dose of 0.12 milligrams per kilogram (mg/kg) compared with 2 mg/kg in dogs without the mutation (Gieseg & Parton, 2012).

Currently, no investigations have been carried out to determine the prevalence of the ABCB1 mutant allele in the New Zealand working dog population, but it is reasonable to assume that this could be an issue as the mutation has been reported in dog breeds known to make up the gene pool for working dogs (including Border Collies, Old English Sheepdogs, Rough-coated Collies, English Shepherds and Australian Shepherds).

Going places — example of secure method for transporting dogs.

Conclusion

Working farm dogs are both an integral and an iconic part of sheep and beef farming in New Zealand. Improvements in the nutrition and health of working dogs will improve their longevity and productivity, and better understanding of training theory will improve their welfare. However, until recently there have been very few studies in this area. The results of recent studies have improved our understanding of working farm dogs and form an important basis for future research.

References

Anon. (2015). Story: farm dogs. Te Ara Encylopaedia of New Zealand. Accessed at www.teara.govt.nz/en/farm-dogs/page-1.

Arnott, E R., Early, J.B., Wade C.M., & McGreevy P.D. (2014). Estimating the economic value of Australian stock herding dogs. *Animal Welfare*, 23(2), 189–97.

Beef & Lamb New Zealand. (2015). Compendium of New Zealand farm facts. Wellington: Beef & Lamb New Zealand.

Cave, N. (2009). Feeding working farm dogs. Wellington: Sheep and Beef Cattle Veterinarians of the New Zealand Veterinary Association.

Cave, N., Allott, S., & Harland, A. (2013). The serological response to a non-adjuvanted, 3-component leptospirosis vaccine in farm dogs. *International Sheep Veterinary Congress*, March, 66. Wellington: New Zealand Veterinary Association.

Cave, N., Bridges, J., Cogger, N., & Farman, R. (2009). A survey of diseases of working farm dogs in New Zealand. *New Zealand Veterinary Journal*, 57(6), 305–12.

Cave, N.J., Harland, A.L., & Allott, S.K. (2014). The serological response of working farm dogs to a vaccine containing *Leptospira interrogans* serovars Copenhageni and Pomona, and *L. borgpetersenii* serovar Hardjo. *New Zealand Veterinary Journal*, 62(2), 87–90.

Cogger, N., Jerram A., & Stevenson, M. (Submitted 2016). A cross-sectional survey of working farm dogs in New Zealand: description of population. *New Zealand Veterinary Journal*.

Dalton, C. (1996). *Farm dogs — breeding, training and welfare*. Auckland: New Zealand Rural Press and Waikato Polytechnic.

Dalton, C., & Orr, M. (2004). *The sheep farming guide*. Christchurch: Hazard Press.

Gieseg, M., & Parton, K. (2012). Do New Zealand farm working dogs carry the MDR1 mutation? *Vetscript*, 25(II), 56.

Guilford, G. (1997). Nutrition of the working dog. Combined seminar, Society of Sheep and Beef Veterinarians and Companion Animal Society NZVA, Vetlearn Foundation, New Zealand.

Hendriks, M., Hill, K., Cogger, N., Jones B., & Cave, N. (2012). A retrospective study of gastric dilatation and gastric dilatation and volvulus in working farm dogs in New Zealand. *New Zealand Veterinary Journal*, 60(3), 165–70.

Hill, K. (2010). Gastric dilatation volvulus in the working dog. *Proceedings of the 3rd AVA/NZVA Pan Pacific Veterinary Conference*. St Leonards, NSW: Australian Veterinary Association.

Hughes, P. (2001). Hip dysplasia in the New Zealand huntaway and heading dog. *New Zealand Veterinary Journal*, 49(4), 138–41.

Hughes, P. (2013). History and development of the farm working dog in New Zealand. *International Sheep Veterinary Congress*, March, 101.

Jerram, A. (2013). Demographics and management of dogs used to herd livestock in New Zealand (Master's thesis). Palmerston North: Massey University.

Ministry for Primary Industries (2010). Controlled area notice in respect to hydatids: Notice 293. Ministry for Primary Industries.

O'Connell, A., Scott, I., Cogger, N., Jones, B., & Hill, K. (2013). Prevalence of gastrointestinal nematode and protozoan parasites, body condition scores of working sheep dogs in New Zealand in 2010. *International Sheep Veterinary Congress*, March, 126.

O'Keefe, J., Jenner, J., Sandifer, N., Antony, A., & Williamson, N. (2002). A sero-survey for antibodies to *Leptospira* in dogs in the lower North Island of New Zealand. *New Zealand Veterinary Journal*, 50(1), 23–25.

Oliver, M., Sheild, T., & Jerram, P. (2004). *I am a working dog*. Dunedin: Longacre Press.

Ovis Management. (n.d.). Safe feeding of meat. Accessed at www.sheepmeasles.co.nz/on-the-farm/safe-feeding-of-meat/.

Parton, K. (2002). Prevalent toxicological agents in the working dog. 32nd Annual Seminar, Society of Sheep and Beef Cattle Veterinarians NZVA, Vetlearn Foundation.

Parton, K., Wiffen, E., Haglund, N., & Cave, N. (2012). Macrocyclic lactone toxicity due to abamectin in farm dogs without the ABCB1 gene mutation. *New Zealand Veterinary Journal*, 60(3), 194–97.

Redwood, M. (1980). *A dog's life. Working dogs in New Zealand*. Wellington: A.H. & A.W. Reed.

Rennie, N. (1984). *Working dog*. Auckland: Shortland Publications.

Reynolds, A., Reinhart, G., Carey, D., Simmerman, D., Frank D., & Kallfelz, F. (1999). Effect of protein intake during training on biochemical and performance variables in sled dogs. *American Journal of Veterinary Research*, 60(7), 789–95.

Sheard, H. (2014). Demographics and health of New Zealand working farm dogs: a survey of dogs on sheep and beef farms in New Zealand in 2009 (Master's thesis). Palmerston North: Massey University.

Singh, I., Tucker, L.A., Gendall, P., Rutherfurd-Markwick K.J., Cline, J., & Thomas, D.G. (2011). Age, breed, sex distribution and nutrition of a population of working farm dogs in New Zealand: results of a cross-sectional study of members of the New Zealand Sheep Dog Trial Association. *New Zealand Veterinary Journal*, 59(3), 133–38.

Subcommittee on Dog and Cat Nutrition. (2006). Nutrient requirements of dogs and cats. Washington DC: National Academies.

Acknowledgements

The editor would like to thank those members of the Institute of Veterinary Animal and Biomedical Sciences of Massey University who helped in many ways with the production of this book.

The science underpinning the production of livestock becomes greater every year, and a short, easy-to-read and simple introduction to livestock production has become more necessary for high-school students studying agriculture. It is also suited to university students, including those involved in programmes such as nutrition, agri-commerce and environmental studies. This primer is also written for those who are interested in food production and animal welfare.

The editor thanks all those who wrote the chapters of this book for their enthusiasm and hard work and for being willing to produce easy-to-read descriptions of how their specific species is produced.

He would also like to thank a few others who assisted with specific chapters: Briar Robinson (goats), Professor Keith Thompson, Dr Nick Sneddon and Associate Professor Richard Laven (dairy cows), and Dr Christine Christensen (beef cattle).

Image credits

Lisa Whitfield: pp2–3, 6–7, 11, 20; **iStock:** pp12–13, 97, 106, 130, 136, 139, 144, 170–71, 250–51, 271, 280–81; **Kristina Mueller:** pp17, 23; **David Wiltshire/ Massey University:** pp24, 26; **Penny Back:** pp28, 37, 38, 41; **DairyNZ:** p32; **Jenquip:** p40; **Sarah Ivey/ Angus Pure NZ:** pp 56–57, 59, 69, 71, 73, 77, 83; **Morrison Farms:** pp65, 80; **Steve Morris:** pp74, 75, 81; **R Hickson:** p79; **Lydia Cranston:** pp84–85, 87, 91, 93, 101, 104, 113, 115, 116, 118, 121; **Peter Wilson:** pp124–25, 127, 128, 137, 140, 141, 143; **Gosia Zobel:** pp146–47, 150, 153, 164, 167; **Keith Thompson:** pp149, 156, 159, 161; **Colin Prosser:** pp151, 152, 154; **I Barugh:** pp173, 181, 183, 184, 187, 212; **Freedom Farms:** pp192, 193, 195, 208, 210; **Poultry Industry Association of New Zealand:** pp214–15, 217, 221, 230, 231, 232, 241, 244, 245, 247; **Charlotte Bolwell:** pp256, 258, 269; **L Gillespie:** p263; **Chris Rogers:** p275; **Helen Sheard:** p283; **Ally Dowle:** p285; **Amy Jerram:** p286; **Helen Williamson/Vetlife:** pp287, 288, 291; **Naomi Cogger:** p295.

About the contributors

Dr Penny Back is senior lecturer in dairy production at Massey University. She specialises in dairy systems and calf management. She grew up on a sheep and beef farm in Taranaki which is now a dairy farm. Penny has a Bachelor in Agricultural Science from Massey and has worked with DairyNZ and the Livestock Improvement Corporation.

Dr Charlotte Bolwell completed a BSc in equine science and an MSc in veterinary epidemiology in the UK, before moving to New Zealand in 2007 to begin a PhD entitled 'Epidemiological studies of early exercise and management in Thoroughbred racehorses'. Charlotte completed her PhD in 2011 and is now a lecturer in equine studies at Massey University. Charlotte's research utilises key aspects of epidemiology to address fundamental questions with respect to the health, production and performance of horses in New Zealand.

Dr Naomi Cogger began her research career investigating injuries in racehorses at the Faculty of Veterinary Science at the University of Sydney. In 2003 she moved to New Zealand and since then her studies have expanded to improving our understanding of health and performance in working farm dogs. She is currently a director of Massey University's Working Dog Centre. Naomi's interest in farm dogs is also personal as she lives on a beef farm and is the part owner of a NZ Heading Cross Smithfield.

Dr Lydia Cranston comes from a farming background and has always been interested in agriculture. She is currently a lecturer in sheep production systems at Massey University and on weekends she works on her own sheep and beef farm. Her main areas of research include the use of alternative feed types to improve ewe and ewe lamb breeding performance and lamb growth pre- and post-weaning.

Dr Erica Gee is a senior lecturer in equine science at Massey University. She completed her veterinary degree at Massey University. After working in large animal practice she returned to Massey to complete a PhD. Erica completed a residency in equine reproduction at Colorado State University, and is a Diplomate of the American College of Theriogenologists.

Dr Andy Greer comes from a farming family in Southland and has maintained an interest in animal production. After his PhD, he took up an AgMARDT fellowship to work in a postdoctoral position at Moredun Research Institute in Scotland before returning to Lincoln in 2008, where he is currently a senior lecturer in animal science, teaching livestock production and nutrition. His main areas of research include gastrointestinal parasitism and sheep production systems.

Kate Griffiths is a Massey University veterinary graduate (BVSc Dist.). She worked as a mixed animal veterinarian in rural New Zealand before returning to Massey University in 2015. Currently she teaches in the veterinary programme with a focus on deer, sheep and beef cattle. Her research interests include productive longevity and wastage in commercial ewes and on-farm ewe production and health.

Professor Paul Kenyon grew up on a sheep and beef farm before heading to university to undertake an agricultural degree. He is currently the head of the Institute of Veterinary, Animal and Biomedical Sciences at Massey University and a Professor of Sheep Husbandry. His research programmes include: maximising ewe breeding performance, twin and triplet lamb survival and growth to weaning, effects of body size on efficiency of production in sheep systems, alternative feed types to improve sheep performance and fetal programming.

Professor Patrick Morel gained both his undergraduate and postgraduate degrees from the Swiss Federal Institute of Technology in Zurich. After his PhD, he worked as a geneticist at the Swiss National Pig Centre, where he set up a new selection index that included intramuscular fat. He was appointed lecturer in pig husbandry at Massey University in 1991. His research has focused on feed evaluation, nutrition, growth, pork quality and modelling nutrients flow in pigs.

Professor Steve Morris is Professor in Animal Science and leader of the Animal Production and Health Group, Animal and Biomedical Sciences at Massey University. Steve is one of New Zealand's most experienced and knowledgeable sheep and beef productivity experts. His recent research continues to investigate how on-farm factors can influence the quality of products from sheep and beef cattle production systems. He has published over 175 refereed journal papers and over 250 other conference and technical papers. He is also Director of the International Sheep Research Centre at Massey University.

Dr Colin Prosser BSc (Hons), PhD is Chief Scientific Officer at Dairy Goat Co-operative (DGC). Colin obtained a BSc in biochemistry and PhD in human lactation from the University of Western Australia. Colin worked in the USA and UK before coming to New Zealand to conduct research into the composition of goat milk and the nutritional and physiological benefits of goat milk formulations. Colin has written 85 peer-reviewed publications related to milk and milk producing animals.

Professor Velmurugu Ravindran is Professor of Poultry Science at Massey University. He graduated with a BSc in agriculture from the University of Sri Lanka and MSc and PhD degrees in animal nutrition from Virginia Tech University, USA. He held academic and research positions in Sri Lanka, Canada, the USA and Australia, prior to moving to New Zealand in 1998. He has had a distinguished international career in nutritional science, having published in excess of 400 scientific works.

Dr Chris Rogers completed a PhD at Massey University in equine biomechanics. He was then awarded a Huygens Post-Doctoral fellowship at Utrecht University, the Netherlands. Upon returning to Massey University Chris has been leading a small dynamic team of researchers focusing on maximising the economic and physiological opportunities of pasture-based equine production. Outside of the university, Chris has an active role in consultancy and administration within equestrian sport.

Helen Sheard graduated from Massey University in 2003 with a BSc in veterinary science and has worked in rural practices since, from Raetihi to Feilding. She became interested in working in dog health after adopting a string of retired huntaways, and has recently completed a Master's degree in veterinary medicine that included a dissertation on working dog health. Due to a recent human addition to the family, the current huntaway inhabitants are limited to two.

Professor Kevin Stafford is Professor of Veterinary Ethology at Massey University. He is a veterinary graduate from Ireland. His research interests include animal behaviour, production and welfare. He has published several books on animal welfare and authored more than 200 refereed papers. He has a small sheep farm.

For more information about our books please visit
www.masseypress.ac.nz

First published in 2017 by Massey University Press
Private Bag 102904, North Shore Mail Centre
Auckland 0745, New Zealand
www.masseypress.ac.nz

Design by Kate Barraclough
Figures drawn by Janet Hunt

A catalogue record for this book is available from the National Library of New Zealand

Printed and bound in China by Everbest

ISBN: 978-0-9941363-1-2